# YELLOWSTONE PLACE NAMES

# Yellowstone National Park

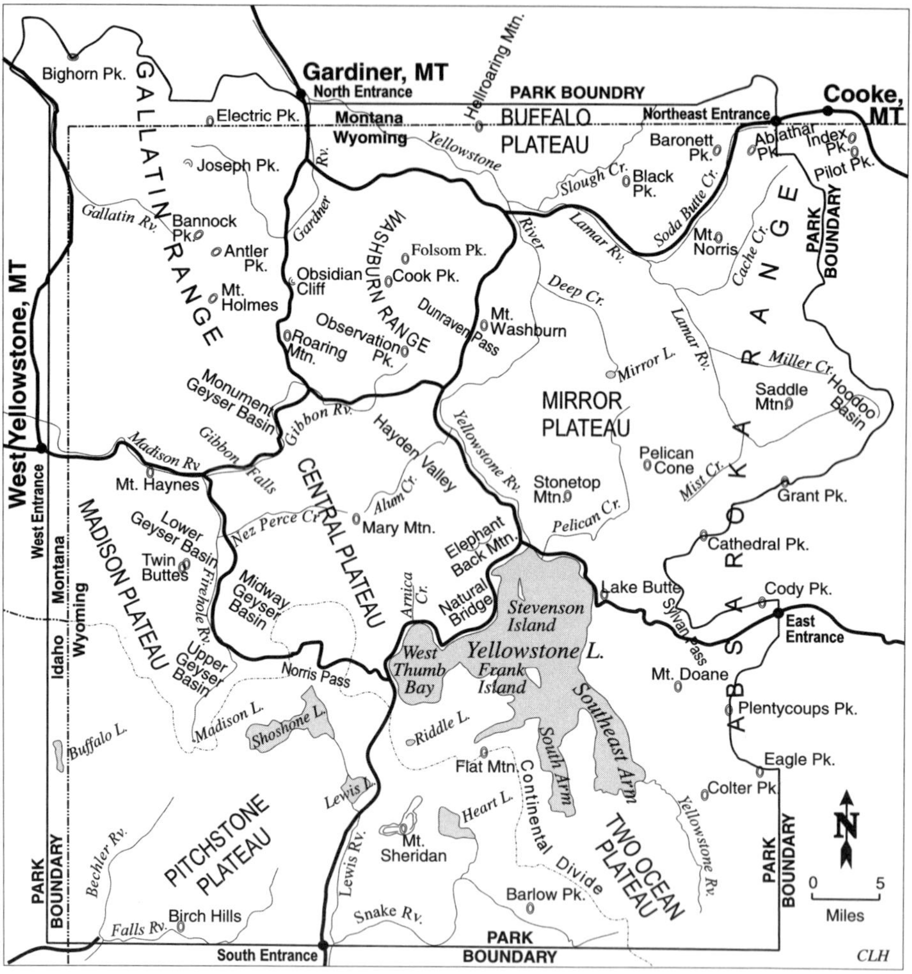

# Yellowstone Place Names
## *Mirrors of History*

*Aubrey L. Haines*

UNIVERSITY PRESS OF COLORADO

Published by the University Press of Colorado
P. O. Box 849
Niwot, Colorado 80544
(303) 530–5337



The University Press of Colorado is a cooperative publishing enterprise supported, in part, by Adams State College, Colorado State University, Fort Lewis College, Mesa State College, Metropolitan State College of Denver, University of Colorado, University of Northern Colorado, University of Southern Colorado, and Western State College of Colorado.

**Library of Congress Cataloging-in-Publication Data**

Haines, Aubrey L.
Yellowstone place names : mirrors of history / Aubrey L. Haines.
p. cm.
ISBN 0-87081-382-x (cloth: alk. paper). — ISBN 0-87081-383-8 (paper: alk. paper)
1. Names, Geographical—Yellowstone National Park. 2. Yellowstone National Park—History, Local. I. Title.
F722.H164 1996
978.7'52—dc20 96-4463
CIP

Unless otherwise noted, all photographs are courtesy of Aubrey L. Haines.

This book was set in Caslon and Gill Sans.

The paper used in this publication meets the minimum requirements of the American National Standard for Information Sciences—Permanence of Paper for Printed Library Materials. ANSI Z39.48–1948
∞

10 9 8 7 6 5 4 3 2 1

This book is dedicated to Wilma,
my helpmate of fifty years.
She is the light of my life.

# CONTENTS

# ILLUSTRATIONS

# INTRODUCTION

Names are applied to the geographic features of a region—to its mountains and valleys, streams and lakes, and to those works of humans superimposed upon the natural terrain—as a means of identification; and through the use of such *place names,* people are able to communicate concerning a locality or an event related to it. Thus, the primary purpose of place names is to ensure identification of the significant features of an area.

But place names can be more than mere "handles" for landmarks and localities. They may also reveal much about the people whose more or less enduring legacy they are. Language, belief, loyalty, accomplishment, humor, or frustration may stand out vividly, hinting of an underlying story that often is a really charming vignette of history or experience, mirroring a so-different past.

It is this second characteristic—the story content of the place names of Yellowstone Park—that is the real burden of this work. Of course, many names appearing on the Park map are devoid of story interest, either from being given casually or because the circumstances are now lost to us, and an attempt to include them all would only result in a meaningless clutter. So, the place names presented here are those relics of the Indian, fur trader, prospector, explorer, administrator, resident, and visitor that have something to say about the discovery and development of our oldest and largest national park.

Before beginning with particular place names, some generalities concerning geographic nomenclature and its application in the Park are in order. Except for the names of political subdivisions, which invariably result from legislation, place names are not bestowed by a government agency but are handed down from the aborigines, explorers, surveyors, early settlers, and later residents of an area, with their permanence depending upon acceptance by the general public (influenced, of course, by mapmakers and writers). This test of "common usage" usually determines the fate of a place name; whether it will survive intact, undergo alteration, or disappear entirely. However, it is not an immutable process, for names relegated to limbo in an earlier time are occasionally resurrected. The early names given to Yellowstone's features gained currency mostly in that manner, with their suitability determined by common usage. But many that appeared after 1882 have arbitrary origins in the dicta of several government agencies and boards. The most important of these—the United States Geographic Board (later renamed the United States Board on Geographic Names)—was established in 1890 as *the* authority for the federal government's use of place names. The board comprises members of several interested agencies, and its decisions regarding spelling, usage, and changes in the place names appearing on official maps and documents are binding upon all branches of the federal establishment. Also the board has provided guidance for state and local agencies interested in geographical nomenclature, and it has promoted public acceptance of uniform standards for naming features.

According to the board, place names are of three types: descriptive, nondescriptive, and mixed. The first, or descriptive type, accents some characteristic of a place or feature. The second type is subdivided into six nondescriptive forms: the *personal* (for important individuals or early residents), the *transplanted geographic* (borrowed from some well-established name of another locality), the *commemorative* (immortalizing some more or less important event), the *religious*, the *casual and whimsical* (folksy names and those of a jocular or fantastic kind), and the *metamorphosed* (in which time and human variability have worked to alter a name until its original form is obscure). The third, or mixed type, includes composites made up of descriptive and nondescriptive elements.

Yellowstone Park has many examples of descriptive place names, such as the Red Mountains, so called from their conspicuous coloration, and the Morning Glory Pool, from its resemblance in form and color to the flower of that name; but the nondescriptive forms are, for all practical purposes, limited to the personal—typified by such prominent examples as Mount Washburn and Peale Island—and to the casual and whimsical, a category that is sometimes represented because of two local peculiarities that will be discussed later. As for the transplanted geographic, the commemorative, and the religious, examples can be found only among such unofficial usages as "Little America" (once applied to an open, windy work camp in Lamar Valley), "Camp Reunion" (the place in the Lower Geyser Basin where two detachments of the Hayden Survey met in 1872 after several months apart), and "St. Mary's Lake" (an entirely mistaken variant of Marys Lake). The single example of the metamorphosed type within the Park is Mol Heron Creek, which takes its avian-sounding name from an early day bartender—one "Friskey John" Mulherin.

The United States Board on Geographic Names does not propose new place names or change those already in use. As William C. Gregg has so aptly noted, "Its general disposition is negative; its mission in life is to prevent inappropriate names and to stop people from putting themselves on the map."[1] However, it does influence the naming of features in all federal areas through its power to pass upon the form, spelling, and appropriateness of the names submitted to it. In its deliberations, the board is guided by the following principles: Euphonious and suitable names of Indian, Spanish, or French origin should be retained; names suggested by peculiarities of features are acceptable, but duplication in such naming should be avoided; names of living persons should be applied very rarely and only in the case of persons of utmost eminence; personal names should not be attached because of relationship or personal interest nor should names of obscure persons be used; names of eminent men now dead, such as early explorers, geologists, and topographers, are generally acceptable; long and clumsy constructions and multiplication of names for different parts of the same feature should always be avoided; in the case of a river, the name should follow up its longest branch, and the use of such terms as fork, prong, or branch in designating tributaries, should be avoided; the spelling and pronunciation should be that sanctioned by local usage; where a word has been changed or corrupted, it is usually advisable to use it so, rather than attempt restoration; wherever a choice is possible, the most appropriate and euphonious should be accepted, and possessive forms and hyphenated spellings are discouraged.[2]

Information supplied by Donald J. Orth of the Board on Geographic Names places the following limitations on the naming of features in national parks and wilderness areas:

> Sec. 7. There is a restriction on the number and kind of names approved for these National areas in an effort to maintain their natural and wilderness character. In general no new names will be adopted for features in a National Park or Wilderness Area unless a need is identified for purposes of safety, education, or area administration. . . . Name proposals commemorating persons are discouraged and will not

normally be adopted unless there is a direct, historical association between a person being honored and the feature.

Sec. 8. The Board on Geographic Names will not adopt a name proposal that implies discrimination or is derogatory to a particular person, race or religion. This also applies to names considered obscene or blasphemous in a present-day cultural context.[3]

Of course, few of Yellowstone's place names came into existence under those ideal strictures. The Indian, the trapper, and the prospector all relied heavily on descriptive nomenclature; however, the trapper also used personal, commemorative, and whimsical forms of nondescriptive naming, while the prospector, who followed him into the area, occasionally gave his whimsy a ribald slant. But deliberate adherence to standards of even the most rudimentary kind did not appear until 1870.

The explorers who entered the area that year under the leadership of Henry D. Washburn, the surveyor-general of Montana Territory, agreed among themselves not to name the features they found after each other. This resolution was ignored on several occasions, yet the names given to Park features were generally quite appropriate.

Doctor Ferdinand V. Hayden's Geological Survey of the Territories spent much of the summers of 1871, 1872, and 1878 in the Yellowstone region, naming many Park features in the course of their extensive scientific investigation and the mapping that accompanied it. Concerning the naming of features, Hayden said:

In attaching names to the many mountain peaks, new streams and other geographical localities, the discovery of which falls to the pleasant lot of the explorer in the untrodden wilds of the West, I have followed the rigid law of priority and given the one by which they have been generally known among the people of the country, whether Whites or Indians. But if, as is often the case, no suitable descriptive name can be secured from the surroundings, a personal one may be attached. Eminent men who have identified themselves with a great cause, either in the fields of science or legislation, come first to mind.[4]

This standard was a cut better than that employed by the Washburn party, yet it too was frequently violated. Priority was often ignored and the men of the survey were prone to name features for wives, sweethearts, friends, and each other, while their widely separated activities led to a duplication of some descriptive names. But again, the overall effect was quite satisfactory.

The United States Geological Survey, which took up the unfinished task of mapping the Yellowstone region after dissolution of the Hayden Survey in 1879, was an organization of professionals who filled in the chinks in the map in a business-like manner. Work in the Park was directed by Professor Arnold Hague, who established an entirely different standard for the naming of Park features by the personnel of his organization:

> In consultation with Dr. Henry Gannett, geologist in charge of geography, it was agreed that the necessary new names to designate the unnamed mountains, valleys and streams should be mainly selected from the beasts, birds, fishes, trees, flowers, and minerals found within the Park or the adjacent country.[5]

This appeared to provide a most satisfactory approach to the problem of naming minor Park features, yet its effect was to litter the map with mediocre and meaningless names. Worse yet, such nondescript place names sometimes supplanted a more colorful and more appropriate nomenclature. The displaced names, which were often shifted to other features through failure to correctly relate them to the terrain, became the bases of very troublesome geographical and historical warpages.

Through all these years, common usage—that decisive arbiter of the fate of place names—operated to perpetuate some usages and suppress others, so that Hiram M. Chittenden recognized the need to devote a section of his book, *The Yellowstone National Park* (1895), to a systematic listing of Park place names in order to introduce a measure of stability. His task was accomplished in an unusual manner, with separate lists of streams, lakes, mountains, and other categories of features. The merit of his compilation lay in the fact that it salvaged much historical information relating to Park place names that might not have been available at a later time.

In 1922 Park Superintendent Horace M. Albright appointed a place name committee consisting of Mrs. Marguerite L. Arnold and Jack E. Haynes to work under his chairmanship in listing existing place names and recommending additions, changes, and deletions. This group functioned through 1928, gradually eliminating much of the confusion that had arisen during more than fifty years of sporadic and mostly uncoordinated naming of Park features. Many desirable changes were made with the intention to restore the original nomenclature and avoid duplication, and the usages were made to conform with the standards of the United States Geographic Board.

The first place name committee ceased to function after Mr. Albright left the Park in 1929, and necessary work along that line was handled by Newell F. Joyner until 1932. Then, Chief Naturalist C. Max Bauer became interested in the Park's place names and eventually began preparation of a manuscript intended for publication. But his work never reached the stage where it could be put into print and it was subsequently relegated to the files of the Park Reference Library.

A second place name committee came into existence at a later date. This committee included Jack E. Haynes, whose long association with Yellowstone Park and interest in all its affairs provided an essential continuity with past work. The principal contribution of this committee lay in coordinating and clearing place name data for the new topographic map the United States Geological Survey was then preparing for the Park, but the desirability of a place name publication was not forgotten.

In 1957 I was asked to revise Doctor Bauer's manuscript for publication by the Yellowstone Library and Museum Association, but the place name committee became inactive following Chief Naturalist David deL. Condon's transfer to another area a year later, and plans for a publication lapsed. With passage of more than three decades, a wealth of additional information has been accumulated, and this—with the legacy of past efforts—is now presented in a form that should make Yellowstone's place names both useful and interesting.

## NOTES

1. Wm. C. Gregg, "The Cascade Corner of Yellowstone Park," *The Outlook,* 129 (Nov. 23, 1921), p. 474.
2. From *Sixth Report of the United States Geographic Board, 1890–1932* (Washington: Government Printing Office, 1933), p. 19.
3. Memorandum, "Proposing Names for Unnamed Domestic Features," by Donald J. Orth, March 8, 1977. Copy forwarded to Yellowstone National Park, November 27, 1979, by the chief, Rocky Mountain Mapping Center, Denver.
4. F. V. Hayden, *Preliminary Report of the United States Geological Survey of Montana and Portions of Adjacent Territories, Being a Fifth Annual Report of Progress* (Washington: Government Printing Office, 1872), p. 8.
5. J. W. Powell, *Eighth Annual Report of the United States Geological Survey, 1886–1887* (Washington: Government Printing Office, 1888), I, p. 152.

# YELLOWSTONE PLACE NAMES

# I

## REMINDERS OF AN INDIAN PAST

*Yellowstone*
*Shoshone/Snake*
*Bannock*
*Indian*
*Dunanda/Gwinna/Tempe/Wahhi*
*Absaroka/Crow*
*Sheepeater*
*Nez Perce*
*Wickiup/Wigwam*
*Miscellany*

Considering that the Yellowstone region was Indian country well into the latter half of the nineteenth century, it is surprising that so few names of Indian origin have found a place on our present maps. Most names now used to identify a few prominent features are only memorials to Indian peoples who were visitors or neighbors, and some names are mere literary accretions; few would be familiar to any of Yellowstone Park's aboriginal inhabitants. Failure of a true Indian nomenclature to survive the period of Indian–white contact is an indication of the tenuous nature of that meeting. The only residents found by white men in what is now the Park were Sheepeaters — impoverished family groups of Shoshonis or Bannocks living furtive, isolated lives upon the frontier of Shoshoni culture and influence — and they had not been there long.

In ancient times, the Yellowstone region was part of an extensive Salishan domain occupied by Flatheads, plains Pend d'Oreilles, and Kutenais. But early in the modern period (now arbitrarily considered as beginning in A.D. 1700 in the trans–Mississippi West) they were displaced by Shoshonis. Those Indians had obtained Spanish horses and some iron weapons from their kinsmen, the Utes, who had in turn obtained them from Comanches following the Pueblo revolt in New Mexico in 1680. Thus equipped for aggressive warfare, they had burst out of their Great Basin homeland to sweep eastward into Wyoming and northward across Montana.

But the Shoshoni triumph was short-lived. In southern Alberta the mounted raiders met Piegans who had obtained guns from Hudson's Bay Company trading posts through Assiniboine and Cree middlemen. With those terrible new weapons, the northern Indians drove the Shoshonis back across Montana, while a Siouan people — the Absarokas, or Crows — dispossessed them of much of their Wyoming conquest.

So it was that the nineteenth century opened upon a territorial balance in which the Yellowstone Plateau formed a corner bastion on a long Shoshoni frontier. It extended from the Wind River Mountains along the Absaroka Range, and then westward across Idaho to the Blue Mountains of Oregon. This frontier was thinly held by Shoshonean have-nots — the Sheepeaters (a mix of Shoshoni "walkers" and Bannock mountain dwellers). At the Yellowstone corner they were exposed to enemies on two sides: to the Piegans of the Blackfoot confederacy on the north, and to the Siouan Crows on the east.

So situated, the Sheepeaters led furtive lives from necessity; their principal protection was the rugged nature of their mountainous home. When white prospectors began to penetrate the Yellowstone Plateau as an outgrowth of the Montana gold rush of the 1860s, the Sheepeaters did not resist the invasion. Instead they slipped quietly away to join their brethren in southwestern Wyoming and central Idaho. With them went all knowledge of whatever names they had for the streams, lakes, and mountains of a land that was theirs for only a century and a half.

There are some place names in Yellowstone National Park that are drawn from Indian sources, but they come down to us through the translations of white men and are no more than reminders of an Indian past.

## YELLOWSTONE

This most important of all the Park's place names is also its oldest — a strong, useful name of remote Indian origin.

Historian Hiram M. Chittenden considered it a translation of the Minnetaree Indian expression, *Mi tse a-da-zi*, which was first rendered in French as *Roche Jaune* and *Pierre Jaune* by traders from Canada.The English equivalent of "Yellow Stone" is attributed to David Thompson of the North West Company in 1798. However, his reference may not be the earliest. John Evans, a Welshman employed by the Spaniards to explore the Missouri River, had produced a manuscript map the previous year showing a "River Yellow Rock" as a tributary of the Missouri. This fact was probably unknown to Thompson.[1]

Chittenden was uncertain as to why those semisedentary river Indians, who lived several hundred miles below the mouth of the Missouri River's great southern tributary, would know it as "the river of yellow rocks." He appears to have been familiar with an earlier statement of the editor of Livingston's first newspaper (the *Livingston Enterprise*) that the river "was so called from the yellow bluffs which mark its lower course" (this on the authority of "those who have lived long among the aboriginal tribes" — presumably frontiersmen with whom the editor was acquainted).[2] While admitting that the Yellowstone is predominantly a river flanked by bluffs of yellow rock, Chittenden was not content that its name originated from the tawny sandstones of its lower course. He looked several hundred miles farther upstream to the colorful Grand Canyon of the Yellowstone as the place of origin.

This attractive idea has been accepted by many people — yet it is probably incorrect. The basis of Chittenden's idea lay in his assumption that the area encompassed by the present Park was known to the Indians who named the river; but that is quite unlikely. The Minnetaree Indians — Siouan Hidatsas — lived in the vicinity of Devils Lake, North Dakota, prior to being forced by pressure from other Indian peoples to the Missouri River in the vicinity of present Bismarck. Thus they had no basis for a traditional knowledge of a region more than six hundred miles upstream from the Mandan towns where they settled. Nor would they have ventured such a distance afoot — they were as yet without horses — into the domain of a different and undoubtedly hostile linguistic group (the Salishans).

Even after the river Indians obtained horses about 1750, the lower reaches of the Yellowstone River — two hundred miles distant — were surely on the periphery of their hunting forays. By then all the country beyond was dominated by the warlike Shoshonis. During later treaty negotiations with the U.S. government, the broadest territorial claims advanced by these river Indians extended only to the mouth of Powder River, which suggests that the upper reaches of the Yellowstone were beyond their geographical knowledge.

The Crow Indians who displaced the Shoshonis along the middle Yellowstone knew that river by an entirely different name. To them it was *E-chee-dick-karsh-ah-shay*, or as rendered in another attempt to put the Crow expression into print, *Encheda-cahchi-ichi*. Both names translate as "Elk River,"[3] a name deriving from the fact that the migration route of elk from their summer range on the Yellowstone highlands to wintering grounds in the lower valleys paralleled the stream. This is confirmed by the French Canadian trader Francois A. Larocque, who added to his 1805 *Journal* some observations on the Crow Indians. He wrote that they inhabited the country at the head of the *River aux Roches Jaunes*, which they knew "by the name of River a la Biche, from the great number of Elks, with which the country along it abounds."[4] Aboriginal use of the name Elk River was also recorded by the English anthropologist William Blackmore, but his entirely different spelling of the Indian expression — *Ca-tan-wa-hove* — is confusing.[5]

In his 1883 discussion of the name Yellowstone, the editor of the *Livingston Enterprise* reminded his readers that mere chance had bestowed that name upon the noble stream flowing past their doors. In his opinion, "had Lewis and Clark reached the head of our valley instead of its furthest extremity, doubtless they would have entered the name in their journal as Elk river and that might have been the name till the present." Even so, it is unlikely that the editor understood just how much a product of chance was the ultimate supremacy of the Yellowstone form of the name.

The U.S. purchase of French Louisiana from a hard-pressed Napoleon Bonaparte was a chancy venture that might well have been contested by Spain or England. Instead, the fortunate acquisition enabled our young nation to field an exploration of the trans–Mississippi West that ultimately popularized the Anglicized Yellowstone name over its French antecedents. Even after the Lewis and Clark Expedition, however, the outcome was a long time in doubt because *Roche Jaune* was so well established with mapmakers that it remained in use into the first quarter of the nineteenth century.

The present form of the name — Yellowstone, rather than the Yellow Stone of David Thompson and William Clark — made its first appearance on the manuscript map of J. S. Daugherty in 1810,[6] but it was not universally used until the end of the fur trade, after 1840.

It does not appear that any use other than naming the river was made of the word Yellowstone until late in the fur trade era. On July 9, 1835, at the crossing of Snake River in Jackson Hole, trapper Osborne Russell noted in his journal: "There was not a man in the party who had ever been at this place or at the Yellow Stone Lake where we intended to go but our leader had received information respecting the route from some person at the Fort and had written the direction on a piece of paper which he carried with him."[7] From that statement it is evident that American trappers had knowledge of Yellowstone Lake and were using that name instead of the original "Eustis Lake" of William Clark's maps. Later it would also be briefly known as "Sublette's Lake." Yellowstone, however, was the name that endured and was officially applied on Washington Hood's 1839 map. Hood, who obtained his information from fur trade sources, also noted "Yellowstone Pass" — present Two Ocean Pass.

Further extension of the use of Yellowstone in designating places in the region came with the mining era. First came "Yellowstone Range" (1863), for the mountains now known as the Absarokas. The name "Yellowstone City" (1864) was given to a brief settlement at the mouth of Emigrant Gulch, downriver from the present Park. Neither of these names persisted.

Yellowstone River and the Lower Falls (308 ft.). Viewed from Artist Point on the east rim of the Grand Canyon of the Yellowstone.

The name "Falls of the Yellowstone" was noted on Captain William F. Raynold's 1860 map, which remained unpublished until after the Civil War. By the time it was published in 1868, the variant form "Yellowstone Falls" appeared in a newspaper article by an ex-prospector writing as LEGH.[8] The same article described the region as the "Yellowstone Hell," an expression sometimes considered to be a frontier put-down of John Colter (though it has not been found in the literature of the fur trade until near the end of that period).

The period of definitive exploration of the Yellowstone region (1869 through 1871) brought no new use of the Yellowstone name, unless Dr. Ferdinand V. Hayden's shifting of "Yellowstone Mountains" from the range east of the Park to the smaller area of present

Two Ocean Plateau can be considered a "new" usage. This alteration soon disappeared.

Establishment of Yellowstone National Park by an act of Congress on March 1, 1872, provided a memorable use of the name Yellowstone, though not immediately. The new Park was born nameless, referred to in the organic act only as "a certain Tract of Land lying near the Headwaters of the Yellowstone River."[9] It remained so until May 10, when Nathaniel P. Langford received a letter from Assistant Secretary of Interior G. B. Cowan advising him of his appointment to the position of superintendent of Yellowstone Park. It should be noted here that there was some interest in a proper name for the park-to-be outside the Congress, if not within, even before the proposed legislation passed the House or Senate. Writing in the Deer Lodge, Montana, *New North-West* on January 13, C. C. Clawson noted that Congress was about to

> set aside 100 square miles here as A WORLD'S PARK, which it no doubt will. And in the scramble for names for this new territory it is to be hoped a suitable one will be found, with no "big Injun" chief figuring in it. Let it be something appropriate, for instance — Hiawatha (High Waters), or Lenawatha (Leaping Waters). Or is that *too* high?"

No, it wasn't too high; just too implausible in its translation to be taken seriously. Another name that was not taken seriously was photographer Henry Bird Calfee's "Yellowstone Rapids," applied in 1878 to the river immediately above the Upper Falls. This use appears to have been limited to the stereoscopic photographs he sold.

Presidential executive orders later extended the area covered by the Yellowstone name to an even larger span of country than the 3,312.5 square miles of the organic act. In 1891, President Benjamin Harrison established the "Yellowstone Park Timberland Reserve," which was subsequently reorganized as the "Yellowstone Park Forest Reserve" and "Yellowstone Forest Reserve."[10]

One more application of the name Yellowstone completes its background. In 1908 a town site came into existence at the terminus of the Union Pacific's "Oregon Short-line" branch to the Park. The new town at the boundary was first called "Riverside, Montana," taking that name from a stage-station site on the Madison River six miles to the east. But the town was not on the river, and there was some confusion in differentiating it from the old site where there still was a station occupied by soldiers guarding the Park. So, the following year Riverside became "Yellowstone, Montana" — creating an even greater problem. Mail intended for Park headquarters tended to end up at the wrong place, and the town's mail often made a detour through Park headquarters at Mammoth Hot Springs. A fully satisfactory solution was not found until 1920, when the troublesome name was amended to read "West Yellowstone."[11]

The foregoing discussion, of course, does not include administrative use of the name in designating the Yellowstone River Trail, which appeared on the Park map in 1959.

## ABSAROKA/CROW

These names are related in that one is an English approximation of the other, although they have a somewhat different connotation. As place names, both date from the period of earliest white knowledge of the Yellowstone region.

Absaroka (pronounced Ab-SOR-kee) is said to have originated in a word used by the Crow Indians to identify themselves. The meaning of the original word, recorded by Captain Clark as *Ap-sah-soo-ha* and by Larocque as *Assarechas,*[12] is given as "children of the large-beaked bird," from the Hidatsu Indian expressions *absa*, "large-beaked bird," and *rokee*, "children" or "offspring."[13]

The Absaroka came to be known as Crow Indians through an accident of sign-language interpretation. Their sign for themselves was a winglike flapping of the arms, which was taken by the Canadian voyageurs to represent the labored flight of the large raven so common to the area. They were thus identified by the French term *Gens de Corbeau*, or people of the raven, and that became in English simply "Crows." This misinterpretation originated prior to 1776, as it appears on Collot's map.

The latter form of the name — Crow — was associated temporarily with the Park through its use in designating the reservation guaranteed to the Indians by treaty in 1868. This Crow Indian Reservation comprised the land lying between the Yellowstone River and the 45th Parallel of North Latitude, westward from the 104th Meridian. As established in 1872, Yellowstone National Park overlapped the reservation by a little more than two miles from the Yellowstone River to the northeast corner of the Park, but that fact seems to have been of no concern until valuable minerals were discovered in the vicinity of

Absaroka Range. As seen across West Thumb Bay of Yellowstone Lake, the mountains outline the eastern boundary of Yellowstone Park. The hot spring in the foreground is Fishing Cone.

present Cooke City, Montana. The resulting clamor for removal of the Indian title from the mining region was seconded by the greedy designs of would-be settlers, ending in cession of the disputed territory by the Crows on April 18, 1882, upon establishment of a new reservation farther east.

With that adjustment of the Crow Indian Reservation, the name of those Indians was also removed from the Park map — but not for long. There is an L-shaped mountain range bordering the Park eastward from the Yellowstone River on the north and roughly approximating its eastern boundary. It was first known as the Yellowstone Range or Yellowstone Mountains, and also as the Snowy Range or Snow Mountains; but Captain William A. Jones, who made what he thought was the first crossing of the range in 1873, presumed to change the name to "Sierra Shoshone."[14]

That euphonious but not very appropriate name was so galling to certain members of the U.S. Geological Survey that they promoted another change — to "Absaroka Range." Historian Hiram M. Chittenden thought this last change, to a name that is almost unpronounceable, was not justified (he preferred Yellowstone Range), but Absaroka has stood.

## SHOSHONE/SNAKE

These names are related in the same manner as the previous pair, and both are products of early white contact with Indians.

The meaning of the name of the Indians now called Shoshonis (the preferred spelling when referring to the people) is lost in the antiquity of the tribe. The best guess traces its origin to an uncomplimentary Siouan expression applied to them by their Crow neighbors. Yet there is no doubt about how they became known generally as "Snakes": it was another of those classic mistakes in sign-language interpretation. The sign used by the Shoshonis to identify themselves was a sinuous motion of the hand with the index finger extended. This was intended to show that they were the grass-weaver people (from their early use of grass mats in erecting habitations), but the sign was misconstrued as the track of a snake; accordingly, they were called *Gens de Serpent.*[15] These "People of the Snake" became simply "Snakes" to the mountain men — no offense intended.

Both forms of the name for these Indians ("Shoshone" is the accepted spelling where place names are concerned) appear on the Park map. Trappers very early came to use the name "Snake" to identify the stream Captain Clark had called Lewis's River, with Lewis's name relegated to that portion above the mouth of Henrys Fork.[16] Trapper Osborne Russell and others, however, knew the main stream that runs through Jackson Hole from the present Park as the Snake River, and that is the way the name has come down to the present.

If James Gemmell is correct, the lake now known as Shoshone may have been called "Snake Lake" late in the fur trade period, though that usage did not persist.[17] Otherwise, the English form of this Indian name has been used only in connection with the naming of a group of hot springs on Snake River and for administrative facilities adjacent to that stream.

The use made of the other form of the Indian name — Shoshone — is more extensive, though also more recent. The attempt of Captain William A. Jones to change the name of the Absaroka Range to Sierra Shoshone has been discussed. Jones may have acquired his liking for the name from the Hayden Survey's use of it in designating a large lake in the southwestern quadrant of the Park. This lake — the same one Gemmell called Snake Lake — had been given the name "DeLacy's Lake" in honor of the prospector-surveyor who first established its relationship to the Snake River drainage. But Professor Frank H. Bradley, chief geologist of the Survey, considered deLacy's map inaccurate and thus was unwilling to allow him the honor.[18] Bradley instead renamed DeLacy's Lake, commenting on the change in the following unkind terms:

> The numerous and outrageous errors of the map show that neither as discoverer nor as mapper of this lake has Mr. DeLacy any claim to a perpetuation of his name; and, since the lake occupies a position entirely different from that assigned to DeLacy's Lake, we have decided to drop that title, and to call this, in our maps and reports, Shoshone Lake, as being the head of one of the principal forks of the Shoshone or Snake River."[19]

Walter Washington deLacy (he preferred the lowercase spelling of his surname) was a mild-mannered man, but he could not overlook such an affront to his professional competence. Because he thought Dr. Hayden was the author of that blunt opinion, he addressed a letter to Major Martin Maginnis, then representing the Territory of Montana in Washington, D.C. He complained, "Profr. Hayden in his published report of 1872 speaks of me in connection with this Lake in very unhandsome & very undeserved terms. He seems on many occasions to have been actuated by unworthy feelings of jealousy." DeLacy added in his own defense that he had never claimed that his map was correct in all particulars, or that he had been entirely over the region in question. "I do claim, however," he stated, "that I was the first to publish the fact that South Snake river took its rise in a large lake, north of Lake Jackson, and flowed into it — a fact confirmed by Dr. Hayden's own surveys."[20]

Although friends of deLacy interceded with Hayden, he did not disavow Bradley's decision beyond a very belated statement that "properly speaking the lake should probably be known as DeLacy's Lake, but, as in the case of Yellowstone Lake, the name that has the right of priority has not been retained."[21] So, deLacy lost his lake. Shoshone Creek, which flows into the lake from the west, and the Shoshone Geyser Basin at the mouth of the creek, were both named by the Hayden Survey from their relationship to the lake.

It must be noted here that Professor Bradley, in giving "Shoshone" as an alternate name for the Snake River, was merely following a usage already established in early romantic literature of the West; it was never popular. Eventually that name was assigned to another stream where it supplanted a more colorful name of remote Indian origin.

That branch of Wyoming's Big Horn River now known as the Shoshone was called "Stinking Cabin River" on William Clark's first map (1805), from information received at the Mandan towns on the Missouri, and it became "Stinking Water" on his later compilation (1811). But it was the form first used by George Drouillard in 1808 — Stinkingwater River — that became the accepted usage during the remainder of the nineteenth century. These were only English approximations of an obscure Indian name that undoubtedly referred to the sulfurous thermal features on the outskirts of present Cody, Wyoming, and others now under the waters of Buffalo Bill Reservoir. With the settlement of that part of Wyoming, such an obnoxious denomination — though appropriate in a way — failed to please most residents.

It fell to Captain Hiram M. Chittenden of the Army Corps of Engineers to initiate the change to Shoshone River. His report on the construction of the East Entrance Road into the Park explains it this way:

> This stream has heretofore been known as the Stinking-water River. It was so named in 1807 by its discoverer, John Colter, who came upon it where there is a large tar spring near the junction of the two forks and gave it the name on account of the impression received at this particular point. The name was wholly inappropriate as the stream is one of the finest in the mountains, both in the purity of its water and the character of its scenery. Local usage has succeeded in changing it to "Shoshone," but this change has never received official recognition. As the valley of the river will soon be followed by a tourist route from upward of 50 miles it was considered important that an appropriate name be adopted, and acting upon a recommendation from this office to the governor of Wyoming, the legislature of that State passed an act (February 14, 1901) changing the name to Shoshone. The stream will hereafter appear in the official reports and maps of the Park under its new name.[22]

With regard to the foregoing statement, Chittenden erred in crediting the name "Stinkingwater" to John Colter. He may have been the white discoverer of the thermal features later called "Colter's Hell," but Captain Clark's use of the name Stinking Cabin River on his 1805 map is evidence enough of his prior knowledge of the stream and its Indian designation.

Three other usages involving Shoshone are known within the Park. One — Shoshone Lake Trail — is administrative, appearing for the first time on the Park map in 1956. The other two are unapproved, though of long standing. A viewpoint on the Old Faithful–West Thumb road from which a part of Shoshone Lake can be seen has been called Shoshone Point ever since 1896. Former scout and ranger Raymond G. Little told me that the men he worked with knew the channel between Lewis and Shoshone lakes as the Shoshone Canal.

## SHEEPEATER

This is an informal name for the only group of Indians who were permanent residents of the Yellowstone Park area in historic times. The name has been attached to several features since they left.

The Sheepeaters were a subgroup within the Shoshoni–Bannock culture and not a "pygmy race" as Superintendent Norris surmised.[23] The Shoshoni homeland — the Great Basin — varied from semidesert to alpine in character, a harsh environment in which life was lived mostly at a subsistence level. Thus, subgroups were recognized on the basis of their dietary staple: "buffalo eaters," "deer eaters," "salmon eaters," "sheep eaters" and "root diggers." The result was a structured society with the buffalo hunters at the top, socially and economically, and the sheep hunters and root diggers at the bottom as have-nots. For each subgroup there was a characteristic lifeway based on particular local conditions.

The Yellowstone region was occupied by scattered family groups eking out an existence in the manner of their Shoshoni and Bannock ancestors of prehorse times — by bow-hunting in the forested glens and on the open ridges frequented by mountain sheep. This together with foraging for berries, fish, grass seeds, and even grasshoppers provided a meager living.

These inoffensive, timid people coexisted with the trappers but were unable to adjust to the incursions of prospectors during the mining era. Following establishment of three reservations in accordance with the unratified treaty of September 24, 1868 — the eastern Shoshoni at Wind River, Wyoming, the Shoshoni-Bannock at Fort Hall, Idaho, and the Bannock at Lemhi, Idaho — the Yellowstone Sheepeaters abandoned their difficult situation and joined their reservation brethren. In the opening years of the new Park, the only evidences of its aboriginal residents were the decaying remains of pole and brush shelters in secluded glens and the driveways and coverts used in the pursuit of quarry.

It was just such traces remaining in the rugged area between Bunsen Peak and the Gardner River that led the Park's second superintendent, Philetus W. Norris, to apply the name Sheepeater Cliffs in 1879.[24] In 1904 the Geological Survey transferred the name to the southeast wall of what later still became Sheepeater Canyon. The steel bridge over Gardner River below the canyon (said to be the highest bridge in Wyoming) has been administratively designated Sheepeater Canyon Bridge. These three features — a cliff, a canyon, and a bridge — are the only names the Park map provides in remembrance of the aboriginal residents.

Before leaving the Sheepeaters, it must be noted that the only aboriginal place name of unquestionable authenticity known for a Park feature (though not on any map) was provided by a Sheepeater Indian in 1873. Major William A. Jones led an expedition to the area that year guided by Togwotee, who had formerly lived south of Yellowstone Lake. He identified the stream now called Chipmunk Creek as *Pa-ah*, but did not give a meaning for that name.[25] There may be another original Indian name in *Git-er-or-guy*, which Langford applied to present Hellroaring Creek — but without saying how he came by it or whether the name was of Sheepeater or Crow origin.[26]

It is possible that a reference to Sheepeater Mountain in the 1888 edition of G. L. Henderson's *Yellowstone Park Manual* refers to present Clagett Butte, but the notice is too vague for certainty.

## BANNOCK

Bannock is the name of a tribe of Idaho Indians who were visitors to the Yellowstone Park area in historic times. Their name is attached to an early trailway across the area. The meaning of the name is uncertain. It derives from their own word — *Panaiti* — which can no longer be translated accurately, but may refer to the characteristic upswept hairdo of the warriors, or to a previous residence farther west (beyond the Blue Mountains of Oregon).

The trailway, which was known as the "Bannack Trail" as early as 1870,[27] was partially mapped under that name by the Hayden Survey in 1878. However, the current usage in Park literature is Bannock Trail, in conformity with the proper spelling of the name of that tribe.

This "great trail of the Indians" was used by Bannocks and others going from the plains of Snake River to hunt buffalo in Montana and Wyoming after 1840. By that time, extermination of the buffalo west of the Continental Divide left those western Indians who had adopted a Plains culture only two alternatives: a subsistence lifeway, as in prehorse times, or long, dangerous hunting forays into areas where buffalo were still plentiful. The Bannocks and some of the Shoshoni with whom they shared the Snake River Plain elected to "go to buffalo."

Their route, which began at the Camas Meadows in Idaho, crossed Targhee Pass into the valley of the Madison River, then climbed over the southern end of the Gallatin Range into the drainage of Gardner River, from which it passed eastward up the Yellowstone and Lamar Rivers to the Absaroka Range. Passes opening upon the lower Yellowstone, the Clarks Fork, and the North Fork of Shoshone River gave access to the buffalo ranges of Montana and Wyoming.

Annual hunting trips including families were made by that two-hundred-mile route up to the early 1870s. The last use of it by the Bannocks was in 1878 when a small band fled eastward through Yellowstone Park at the end of the Bannock War.

Within the Park the Bannock name identifies two features: the Bannock Ford, where the trail crossed Yellowstone River near Tower Fall, and Bannock Peak in the Gallatin Range. The crossing probably served a system of aboriginal trailways much older than the Bannock Trail, and there is reason to believe it is the crossing used by John Colter on his wilderness trek in 1807–1808. It is the only place the river can be crossed by a man

Bannock Ford (at the small, upper island). An aboriginal crossing of Yellowstone River near Tower Fall named for the Bannock Indians who used it traveling to and from buffalo ranges east of the present park.

traveling afoot — and then only in winter when the water is low. Also, the notation "Hot Spring Brimstone" shown on Clark's map is an adequate description of "The Sulphur Beds" — an odoriferous thermal area on the river's east bank.[28]

Bannock Peak, named by the Geological Survey in 1883, has no direct historical connection with those Indians. Another Bannock Indian name — Tendoy Falls, for a band chief of that tribe (given by William C. Gregg in 1921) — honored a man with no direct historical connection with the Park.

## NEZ PERCE

Nez Perce is the tribal name of several bands of Idaho and Oregon Indians who passed through Yellowstone Park in the course of a fighting retreat from their homeland in 1877.

The name is French Canadian in origin and was given the tribe by David Thompson's voyageurs for the same reason that Lewis and Clark called them "Pierced Noses" — because some wore decorative pieces of seashell through the septum of the nose.[29] The pronunciation is distinctively Western: say "Nez Purse." It is an unfortunate appelation, being neither accurate nor resembling their own term (their name

Nez Perce Creek. This stream flows westward from Mary Mountain through a series of meadows to Firehole River. The embattled Nez Perce Indians retreated up it in 1877, pursued by the troops of Gen. Oliver O. Howard.

for themselves was *Ne-Mee-Poom*, "The People," but common usage seems to have decided against that).

Within the Park, their name is associated with the stream they ascended to reach the interior, and the ford at which they crossed the Yellowstone River. Also, a peak is named for a band chief, Young Joseph, who was a prominent figure in the Nez Perce War (1877).

Nez Perce Creek was originally known as the "East Fork of Firehole River," but the cumbersome nature of that early designation did not have the approval of the Geological Survey. Thus, in 1883 the name was changed to commemorate the Nez Perce Indians and their valiant retreat.

Administrative use has added the name Nez Perce Patrol Cabin to the Park map approximately a mile above the mouth of that stream.

Nez Perce Ford is the name given by Superintendent P. W. Norris in 1880 to the place where the Nez Perce Indians crossed Yellowstone River on their retreat through the Park in 1877. This aboriginal ford, which is a mile above the Mud Volcano area, was the only practicable crossing between the outlet of Yellowstone Lake and the Upper Falls; thus it was important in the administration of the Park prior to the building of bridges at those points.

The death of nine buffalo at this place in February, 1946, as they were attempting to cross the river on the ice, led to the informal use of the name "Buffalo Ford" for this ancient crossing. Such usage was particularly regrettable as the Park already had an approved Buffalo Ford just above the mouth of Lamar River. Action by the U.S. Board on Geographic Names restored the older name, Nez Perce Ford, in 1981.

Joseph Peak was the name applied to a summit in the Gallatin Range in 1885 to honor Young Joseph, Chief of the Wallowa band of Nez Perce Indians. He had come to public notice as the chief who surrendered the hostile remnant at the Bear Paw Battlefield, thereby gaining a reputation as a "red Napoleon," though credit for the military successes of the Nez Perces really belonged to the tribal council and to the valor of the individual warriors. But, as a persistent guardian of his unfortunate people — particularly after the defeat — Joseph was a truly great man.

## INDIAN

This generic term forms a part of several of the Park's place names.

Indian Creek was named in 1878 by a Hayden Survey party because of the prominent travois marks of the Bannock Trail descending its north bank, and the stream name has been used to identify the public campground near its junction with Gardner River.

Indian Pond made its appearance as the name of the small lake east of Pelican Creek in the 1881 *Annual Report* of Superintendent P. W. Norris. Of this place, he says: "My favorite camp on the Yellowstone Lake (and it has evidently been a favorite one for the Indian) has ever been upon the grove-dotted bluff, elevated thirty or forty feet above the lake, directly fronting Indian Pond."[30]

The Geological Survey's 1930 edition of the Park-wide topographic map changed the pond's name to Squaw Lake for reasons that remain unclear. That name, which is derogatory to Indian people, remained in use until 1981, when a decision of the Board on Geographic Names restored the early form — Indian Pond.

## WICKIUP/ WIGWAM

Both of these words derive from a proto-Algonquin Indian root indicating a dwelling. The first is more often used to refer to tepee-like structures of poles and brush. The second usually refers to a larger dwelling with an arched frame covered with bark, hides, or mats.

Wigwam Creek was the early name of a minor drainage flowing into Gallatin River near Black Butte. A 1937 decision of the Board on Geographic Names changed the name to Wickiup Creek. Both names

Crow Indian hunting lodges on Wickiup Creek. Such structures, made windproof with hide, blanket, or brush covering, provided overnight shelter for hunting or raiding Indians.

were used because of the presence of a fine group of Crow hunting lodges in that drainage. While often identified as "Sheepeater wickiups," such tepee-like shelters are in the Siouan tradition; Sheepeater dwellings were generally built in lean-to fashion against the face of a cliff or boulder in order to take advantage of reflected heat from a fire at the base of the rock.

A somewhat related term — lodge pole, referring to one of the sticks supporting an Indian tepee — was the Hayden Survey's choice in 1871 for the stream prospectors already knew as Cache Creek. Fortunately, the earlier prospector name has survived.

## DUNANDA/GWINNA/TEMPE/WAHHI

These presumably Indian names appearing on the present Park map are said to be of Bannock origin, though they have not been certified as such.

The need for irrigation water to support the World War I agricultural development in Idaho's Snake River Valley led to attempts to get access to Park waters. A particular proposal of the postwar years was creation of a large storage reservoir in the southwest corner of the Park — its Cascade Corner. The National Parks Association recognized the project as a threat to the integrity of the National Park System and fielded a party to investigate and publicize the scenic values of that area. A member of this 1921 party, William C. Gregg of New Jersey, named a number of previously undesignated features using Indian names he was familiar with.

A waterfall on Boundary Creek was called Dunanda Falls for the sheer, clean drop of its waters. Two other falls were named on the Ferris Fork of Bechler River — Gwinna for a word meaning "eagle," and Wahhi for the double nature of the falls. Tempe Cascade, on Littles Fork, received its name from the associated cavern. The source of these names is not given by Gregg.[31]

## MISCELLANY

Hiawatha and Lenawatha have been mentioned as facetious 1872 proposals for a Park name, but the former was also seriously applied to a thermal feature in the Lower Geyser Basin by G. L. Henderson in 1891. Whatever it was and exactly where can no longer be determined, which is just as well. Another Henderson name — Minnehaha Terrace — was formerly applied to part of the Angel Terrace at Mammoth

Hot Springs. Both are from the poetry of Henry Wadsworth Longfellow (*The Song of Hiawatha*).

One more name of possible aboriginal origin, *Cut-tuh-o'-gwa*, is given as a Shoshoni name for the Gallatin River (this rests entirely upon an unverified newspaper source).[32] There is no way of knowing whether or not it was used to designate that part of the stream within the Park (Indian usage often applied different names to distinct sections of rivers).

## NOTES

1. Hiram M. Chittenden, *The Yellowstone National Park* (Cincinnati: Robert Clarke Co., 1895), pp. 1–2 & 5–7. The stream made its debut, cartographically speaking, as "Rock or Crow River" on Collot's map (1776), but Evans's map (1797) was the first to approximate the modern usage. See Coe No. 303-iii, Beinecke Library, Yale University, New Haven, CT.
2. "The Name Yellowstone," *Livingston* (MT) *Enterprise*, October 23, 1883, p. 1. The editor's statement is confirmed in Nathaniel P. Langford's description of the Yellowstone River, as given in his manuscript lecture notes prepared in 1871. There he wrote: "For part of the distance it is bordered on either side by parallel mountain ranges, and another portion is traversed by a precipitous ridge of yellow sandstone (from which the river receives its name) varying in height from eighty to three hundred feet." Pp. 30–31. Original in Yellowstone Park Reference Library.
3. Don Moyle, "Crow Agency Man Relates Custer Scout's Story of Battle of Little Big Horn," *The Billings* (MT) *Gazette*, June 21, 1953, Sec. 2, pp. 1 & 13. See also, Note 2.
4. François A. Larocque, "A Few Observations on the Rocky Mountains Indians with Whom I Passed the Summer," *The Oregon Country Under The Union Jack*, B.C. Payette, ed. (Montreal, Canada: Payette Radio Limited, 1962), p. 545. It seems probable that Larocque would have written the name "Riviere a la Biche," but I have used the published form.
5. William Blackmore, "Diary — 1872 — Fourth Visit to the United States," Transcription by Brayer, Western Range Cattle Industry Study, State Museum, Denver, Col., 1945. No. 6, p. 113.
6. J. S. Daugherty, untitled sketch map of the Missouri River headwaters from below the mouth of Cheyenne River, 1810, National Archives, RG 92.
7. Osborne Russell, *Journal of a Trapper*, ed. by Aubrey L. Haines (Portland: Oregon Historical Society, 1955), p. 20.
8. Legh R. Freeman, "California," *The Frontier Index*, Laramie City, D.T., May 5, 1868, p. 1.
9. *United States Statutes at Large*, Vol. 17, Chap. 24, pp. 32–33.
10. Benjamin Harrison, Executive Orders of March 30 and September 10, 1891; and Theodore Roosevelt, May 22, 1902.
11. Victor Gondos, Letter to Jack E. Haynes, March 7, 1955, transmitting information from the records of the Post Office Department stored in National Archives, Washington, D.C. See also Chapter 11, *Riverside/Yellowstone*.

12. Entered as the name of a tributary of Clarks Fork of the Yellowstone — present Rock Creek — in the vicinity of Red Lodge, Montana. See "Map of Part of the Continent of North America . . . ," Coe No. 303-iv, Beinecke Library, Yale University. An engraving from this source by Samuel Lewis was published in Nicholas Biddle's *History of the Expedition Under the Command of Captains Lewis and Clark* (Philadelphia, 1814). See also, Larocque, Note 4.
13. Roberta C. Cheney, *Names on the Face of Montana* (Missoula: University of Montana, 1971), p. 1.
14. William A. Jones, *Report Upon the Reconnaissance of North-Western Wyoming, Including Yellowstone National Park, Made in the Summer of 1873* (Washington: Government Printing Office, 1875), p. 21 & map.
15. For this and other place names in Canadian French, see the map that is Coe No. 303-iii, Beinecke Library, Yale University, New Haven, CT.
16. See the map prepared by Washington Hood from fur trade informants (1839), National Archives, RG 77, US 110. Also, the famous "Niles Register" letter written by trapper Daniel T. Potts on July 8, 1827, and first published in the *Philadelphia Gazette and Daily Advertizer*, September 27, 1827.
17. Olin D. Wheeler, ed., "The Late James Gemmell," *Contributions to the Historical Society of Montana*, II (Helena, MT: State Publishing Co., 1896), pp. 331–336.
18. Walter W. deLacy, *Map of the Territory of Montana and Portions of Adjoining Territories . . . 1865,* published variously. The original manuscript is in the collection of the Montana Historical Society, Helena (Case 2, drawer 2). This map went through many published editions and that of 1870, which was greatly improved by the addition of information supplied by the Cook-Folsom-Peterson Expedition of 1869, was a good map for the time.
19. Ferdinand V. Hayden, *Sixth Annual Report of the United States Geological Survey of the Territories* (Washington: Government Printing Office, 1873), p. 244.
20. Walter W. deLacy, Letter to Martin Maginnis, May 9, 1874. A copy is in the collection of the Montana Historical Society, Helena.
21. Ferdinand V. Hayden, *Twelfth Annual Report of the United States Geological Survey of the Territories*, 2 (Washington: Government Printing Office, 1883), p. 249.
22. Hiram M. Chittenden, "Annual Reports upon the Construction, Repair, and Maintenance of Roads and Bridges in the Yellowstone National Park . . . ," *Annual Report of the Chief of Engineers for 1901* (Washington: Government Printing Office, 1901), FFF, p. 3781.
23. Philetus W. Norris, *Annual Report of the Superintendent of the Yellowstone National Park to the Secretary of the Interior for the Year 1880* (Washington, D.C.: Government Printing Office, 1881), p. 35.
24. Ibid.
25. Jones, *Report*, p. 39 & map; see Note 14 for citation.
26. Nathaniel P. Langford, "Notes of a Lecture Given by N. P. Langford During the Winter of 1870–71," handwritten in ink in a notebook on 185 pages. Yellowstone Park Reference Library, p. 69.
27. Louis C. Cramton, *Early History of Yellowstone National Park and Its Relationship to National Park Policy* (Washington: Government Printing Office, 1932), p. 119. See also, Warren C. Gillette, "Diary of Warren Caleb Gillette from August 17, 1870 to September 27, 1870," MS diary in the collection of the Montana Historical Society, Helena, p. 66.

28. The manuscript map compiled by William Clark from informants during the period 1806–1811; see Note 12 for citation.
29. Alvin M. Josephy, Jr., *The Nez Perce Indians and the Opening of the Northwest* (New Haven & London: Yale University Press, 1965), pp. 645–646.
30. Norris, *Report . . . 1880*, (1881), p. 17; see Note 23 for citation.
31. William C. Gregg, "The Cascade Corner of Yellowstone Park," *The Outlook*, 129 (Nov. 23, 1921), pp. 469–476.
32. *Bozeman* (MT) *Chronicle*, Feb. 15, 1970.

# 2

## RELICS OF THE FUR TRADE

| | |
|---|---|
| *Hot Spring Brimstone* | *Gardner/Gardiner* |
| *Burnt Hole/Firehole* | *Bridger* |
| *Elephant Back* | *Two Ocean/Two Water* |
| *Hart/Heart* | *Colter/Coulter* |
| *Sulphur/Sulpher* | *Miscellany* |

Place names traceable directly to the fur trade era are scarce in Yellowstone Park; but not because the men of that rough business lacked geographical knowledge. They knew the mountains and valleys, streams and lakes of their trapping-grounds well, and had names for the major features, yet so few were literate that much of their extensive nomenclature was lost in the hiatus between the mountain men and the prospectors of the succeeding mining era. There remain but a few relics of a colorful time.

American trappers entered the headwaters of the Missouri River on the heels of the Lewis and Clark Expedition sent out by President Jefferson to examine the Louisiana Territory our government had purchased from France in 1803. A member of the expedition — John Colter — was given his discharge at the Mandan towns on the return journey, in order that he might join two of those trappers in a venture on the middle course of the Yellowstone River. As he was leaving the mountains the following year, 1807, he was enticed into the employment of the Missouri Fur Trading Company, and, while its trading post was under construction at the mouth of the Big Horn River, Colter began a winter trip of more than 500 miles in search of Indian customers. In the course of that epic journey he became the first white man to enter what is now Yellowstone National Park, where he saw at least some of its thermal wonders.

Prior to the War of 1812, American trappers depended upon a system of fixed trading posts, laboriously supplied by means of river transportation, and the scope of their activity was always limited by dependence on points of resupply. But upon resumption of trapping by Americans a decade after that struggle, a new method involving an annual rendezvous with supply caravans nearer the trapping-grounds opened the entire northern Rocky Mountains to their activity — the Yellowstone wilderness included.

The presence of American trappers at Yellowstone Lake in 1826 is known from a letter written by one of them, Daniel T. Potts, the following summer, and it is likely that the Yellowstone Plateau was visited by trapping parties nearly every year until the end of the fur trade as an organized American enterprise about 1840.

During those fifteen years when the fur trade flourished such notables as Jim Bridger, Johnson Gardner, Joe Meek, Tom Fitzpatrick, and Jedediah S. Smith explored and trapped in the Yellowstone; while a few lesser, but literate, men — Daniel T. Potts, Osborne Russell and Warren Angus Ferris — who were also there, recorded something of current trapper nomenclature in their accounts. Beyond that, we have only the secondhand evidence of those map-makers and explorers whose geographical knowledge profited from the fur trade.

## HOT SPRING BRIMSTONE

The earliest place name in the Yellowstone Park area, and one of historical importance even though it long ago disappeared from current maps, is that curious notation on William Clark's map at the point where John Colter forded the Yellowstone River on his winter trek across the Yellowstone Plateau.[1] Its significance stems from the fact that it is very good evidence Colter really did pass through the present Park in 1807–1808, and is thus entitled to be called its white discoverer.

As mentioned in Chapter 1, in connection with the Bannock Ford, this crossing of the Yellowstone River just above the mouth of Tower Creek is the only place below the falls of the Yellowstone where a man might cross afoot, and then only during the low water of late fall or winter. Also, the east bank of the river immediately above the crossing is lined with fumaroles known locally as the "Sulphur Beds," which accounts for the impression of brimstone. There are some hot springs on the west bank at the mouth of Tower Creek, but they are not as noisome as the Sulphur Beds.

The Bannock Ford is not only unique in its odoriferous thermal area, it also gave Colter access to an Indian trail which undoubtedly existed long before the Bannocks made use of it, and that aboriginal track pointed him through the mountains by an easy pass, which now bears his name, to Clarks Fork of the Yellowstone and country with which he was already acquainted.

A Lewis and Clark place name, added to the Clark manuscript map as Eustis Lake, was also based on information obtained by John Colter during his winter journey. It honored Secretary of War William Eustis; however, that name was never popular, though used on a number of early maps, and trappers were identifying the lake with its nourishing river, the Yellowstone, before the end of the fur trade.

## GARDNER/GARDINER

Both forms of this name are from the same source — the surname of an American free-trapper who plied his trade in a circular mountain valley, or "hole," on the headwaters of Gardner River in 1831–1832. Of all the Park names in current use, Gardner is second only to Yellowstone in its antiquity.

The man immortalized through the use of this name was one of the harder characters engaged in the fur trade. He probably entered that business in 1822 or 1823 as a trapper with the Ashley-Henry enterprise in the upper Missouri country, but nothing is certainly known about him until 1824, when he appeared on Bear River with a band of free trappers. A year later he achieved notoriety by despoiling Peter Skene Ogden, leader of a Hudson's Bay Company brigade (accomplished by buying-up furs Ogden's trappers already owed the British company).

Gardner remained in the northern Rocky Mountains until 1830, when he shifted his base of operations to Fort Union, the American Fur Company post opposite the mouth of the Yellowstone River. He appeared on the account book there on July 12, 1831, through the sale of fifty-three beaver skins and one otter for $347.[2] A trapping agreement entered into with Kenneth McKenzie a year later indicates that Johnson Gardner was an illiterate as the instrument was signed with an "X, his mark." Among the things agreed to was the price of furs Gardner then had "en cache on the Yellowstone River," so that it must have been the fall of 1830 or spring of 1831 when he first trapped the beautiful mountain valley known to the trappers as "Gardner's Hole."[3] The following winter Arikara Indians killed old Hugh Glass on the ice of the Missouri River, afterwards coming into Gardner's camp with some of the trapper's personal property. Several of the Indians were seized, scalped and probably burned alive, and frontier gossip has it that the Arikaras served Gardner the same way at a later time.

The Gardner River, which drains the eastern flank of the Gallatin Range where it projects into the Park, made its appearance as "Gardner's Fork," and it was always referred to that way, or as "Gardner's River," during the fur trade era.[4] That is in agreement with the spelling of his name as it appears in the Fort Union account books.

The prospectors who entered the Yellowstone region during the Montana gold rush of the sixties seem to have been unaware that Gardner River had a name of trapper origin (understandable in the light of the hiatus of more than a decade between the two periods). So, they renamed the river, from the large volume of hot water entering from Mammoth Hot Springs, as "Warm Spring Creek" and "Warm Stream Creek."[5]

The practice of spelling Gardner's name with an "i" — as Gardiner — appears to have resulted from a phonetic error at the time of the Washburn expedition into the Yellowstone region in 1870. Historian Chittenden, while puzzling over the origin of the Gardiner name in 1895 (and yet unaware of the activity of trapper Johnson Gardner in the Yellowstone region), thought it "reasonable to suppose that this name came from an old hunter in the early years of the century, and that the Washburn Party received it from some surviving descendant

Gardners Hole on the headwaters of Gardner River. Second oldest place name in Yellowstone Park, it remembers Johnston Gardner, an American who trapped beaver here in 1832.

of those times."[6] It was an excellent surmise which was verified when Nathaniel P. Langford published his *Diary* in 1905.

The informant was mountain man Jim Bridger, an employee of the wagon road company organized by Langford in 1866. Bridger is credited with accurately describing the eruption of a Yellowstone geyser,[7] and thereby convincing Langford of the existence of the Yellowstone "wonders." Bridger was familiar with the trapper use of Gardner's name to identify an important tributary of Yellowstone River (it was part of the information he provided DeSmet for the map prepared at the Fort Laramie treaty ground in 1851), and, significantly, it reappeared, after long disuse, in the writing of this man who discussed the Yellowstone region with Jim Bridger. Since Jim was a Virginian, I believe his drawl rendered Gardner as "Gah-din-ah"; anyhow, Langford was the first to write the name with an "i" in it, and other members of the Washburn expedition did the same.

The publicity generated by the Washburn expedition in the form of newspaper and magazine articles had the effect of establishing Gardiner as common usage, and the United States Geological Survey gave its official sanction by using the name that way on the topographic map produced in 1885 (despite consistent use of Gardner on the prior maps of the General Land Office and the United States Geological Survey of the Territories). Following World War II, the map of Yellowstone Park was redrawn by modern photogrammetric methods, and, in the course of that work, the Geological Survey returned the spelling to Gardner on all new maps. That usage, which is in agreement with historical facts not known earlier, now prevails except in the case of the name of the town of Gardiner, Montana.

Local legend persists in attributing the town name to a "Jim Gardiner," or, in Mrs. Edith Ritchie's version, to "the Gardiner brothers." Of the latter, nothing confirming their existence has been found in any public record, including the register of graves in the "Tinker's Cemetery" where they were supposedly buried,[8] but Jim Gardiner *can* be accounted for.

Establishment of a settlement on the north boundary of Yellowstone Park at the mouth of Gardner River was a direct result of the eviction of an unauthorized settler from his claim at Mammoth Hot Springs in 1879. James C. McCartney — "Jim" to all his friends — obeyed the superintendent's order reluctantly, so much so that he didn't get quite out of the Park! The desert land claim he took on the bench above the mouth of Gardner River (spelled then with an "i") straddled the Park boundary, as later surveyed. There he built a cabin, saloon and barn just inside the present Park fence, and a race track where the old hay meadow now is.

McCartney's place soon became the focal point for the settlers outside the Park, and, in those days of limited, informal communication, it became the practice to consign mail and merchandise sent to residents in that vicinity to "Jim on the Gardiner," and that was soon reduced to just "Gardiner." A post office using that name was established there on February 9, 1880,[9] and the beginnings of a rude settlement followed.

A fortunate reminiscence of S. P. Panton, a Northern Pacific Railway location engineer engaged in surveying the Park Branch line in 1882, establishes Jim Gardiner and James C. McCartney as the same person. He says: "We met Jim Gardiner at his settlement. He told us that he had located a homestead of 160 acres, including Mammoth Hot Springs, before the park was thought of, and had since been driven out of the park. But he was suing the government for return of his remarkable homestead and expected to win it."[10]

In addition to the forms of the Gardner/Gardiner usages already mentioned, several others must be noted. Present Osprey Falls was earlier known as Gardiner River Falls, and the Gardiner River Bridge of early accounts should not be confused with the present Sheepeater Canyon Bridge which replaced it in 1939 (the old steel bridge was a quarter mile nearer the mouth of Sheepeater Canyon). Also, old records refer to the "Gardiner Addition," a 7,600 acre tract of land which Congress authorized to be added to Yellowstone Park by act of May 26, 1926. This land was considered essential winter range for antelope, and it was formally annexed to the Park October 20, 1932, on receipt of deeds to private lands purchased by the Gallatin Game Preservation Company.

## BURNT HOLE/FIREHOLE

These names, though they appear to represent variations on a common theme, really do not; the one, as first used, was descriptive of a forested valley devastated by fire, while the other is a change — in form, idea and locale — which has created a metaphor to explain subterranean forces.

Burnt Hole is the older of the two names. It was used by the trappers, at least as early as 1833, to describe the Madison Valley. Warren Angus Ferris, who travelled in the vicinity in May of 1834 as an employee of the American Fur Company, notes: "The Burnt Hole is a district on the north side of the Piney Woods [Madison Plateau] which was observed to be wrapped in flames a few years since. The conflagration that occasioned this name must have been of great extent, and large forests of half-consumed pines still evidence the ravages."[11]

Trapper Osborne Russell, who visited the Yellowstone Plateau a number of times from 1834 to 1840, described the Burnt Hole on August 31, 1835, as "a prarie Valley about 80 mls in circumference surrouned [Sic] by low spurs of pine covered mountains which are the sources of a great number of streams which by uniting in this valley form the Madison fork." He also located the Burnt Hole with relation to the Lower Geyser Basin on July 16, 1839. As his party turned up Nez Perce Creek, he wrote: "From where we left the Main fork [Firehole River] it runs in a NW direction about 40 Mls before reaching the Burnt hole."[12]

The distance is closer to 30 miles, but the direction is correct and leaves no doubt that the present Madison Valley was meant. Captain William F. Raynolds, of the Army's Corps of Topographical Engineers, also had an opportunity to locate the Burnt Hole while carrying out his orders to explore the headwaters of the Yellowstone River in 1859–1860. A four-man party left the line of march at the outlet of Henrys Lake to visit Targhee Pass, with the following entry going into the official report:

After crossing Lake fork, Mr. Hutton, Dr. Hayden, and two attendants turned to the east and visited the pass over the mountains, leading into the Burnt Hole valley. They found the summit distant only about five miles from our route, and report the pass as in all respects equal to that through which the train had gone [Raynolds Pass]. From it they could see a second pass upon the other side of the valley, which Bridger states to lead to the Gallatin. He also says that between that point and the Yellowstone there are no mountains to be crossed.[13]

Thus, two quite reliable trappers and an army officer who had Jim Bridger for his guide identified the Burnt Hole with the present Madison Valley. The translocation of the name to a less extensive area — the geyser basins on the Firehole River — was encouraged by Captain Washington Hood's inexact rendering of the spatial relationships of Yellowstone's topographic features. His information was also from fur trade sources, but secondhandedly so; with the result that he showed the Burnt Hole much too large and crowded it too close to Yellowstone Lake.[14] By this juxtaposition no room was left for showing the geyser region west of the lake. When the far-ranging Jesuit missionary, Father Pierre J. DeSmet, sat down with Jim Bridger at the Fort Laramie treaty council in 1851, he obtained information which appears on two maps. One was drawn by Bridger and captioned by DeSmet (Jim was an illiterate), and the other is DeSmet's own refined plat which includes information not on Bridger's map.[15] Bridger's hydrography shows a Madison River which is well-placed but lacking in minor details, while DeSmet's revision has a branch entering from the south and labeled "DeSmets L. and Riv.," which is about right for the upper course of the Firehole River from its source in Madison Lake, and another joining from the east labeled "Fire Hole Riv." (in the place of present Nez Perce Creek). Upon what authority DeSmet added his information, and in particular, the name "Fire Hole," is unknown, though it is tempting to ascribe it to a misunderstanding of something Bridger told him about the Burnt Hole.

Walter W. deLacy, a prospector with a background in civil engineering, produced his *Map of the Territory of Montana and Portions of Adjoining Territories* in 1865, and it was widely used on the mining frontier.[16] This map showed a stream labeled "Fire Hole" entering Madison River from the east, just as DeSmet's map had.

DeLacy was later employed in the Helena office of the Surveyor General of Montana, and it is likely he was the compiler of the manuscript map Henry D. Washburn submitted to the Commissioner of the General Land Office November 1, 1869. The Yellowstone region is portrayed essentially as on the 1865 deLacy map, including a Fire Hole River.[17] Thus, the Washburn party of explorers who entered the Yellowstone region in 1870 had knowledge of that place name through their leader, if not from common usage.

The Washburn party was consistent in the use of the name Fire Hole as it was applied to the river, using it to denote the entire headwater of the Madison River above the present Madison Junction.[18] The stream was referred to in that manner by three of the five diarists with the party, and also the three correspondents who furnished articles to newspapers. It is worth noting that Lieutenant Doane's official report also implies that the "Fire Hole" was really the geyser basins on that river. He wrote: "This valley is known in the wretched nomenclature of this region as the Fire Hole, and contains phenomena of thermal springs unparalleled upon the surface of the globe."[19]

Doane's evident disdain for this name — Fire Hole — is not explained, but it may have had a basis similar to Nathaniel P. Langford's dislike of the satanic nomenclature used by the party. (*See* Chapter 4, Note 35).

The modern form of this name — Firehole — appeared on several maps of the United States Geological Survey of the Territories (Hayden Survey) in 1878, and that was the form used by the later United States Geological Survey on all its maps.

Firehole has been a popular place name, with the result that there are now a number of well-established variant forms. These are: Firehole Falls (1895), in the canyon immediately above Madison Junction; Firehole Spring (1896), a name applied by F. Jay Haynes to some thermal feature in the Lower Geyser Basin and re-affixed, hopefully to the right feature, in 1922; Firehole Lake (1904), a body of hot water on the east side of the Lower Geyser Basin, and Firehole Canyon (1928), the canyon above Madison Junction. Prior to the opening of Fountain Hotel there was a Firehole Hotel at the mouth of Nez Perce Creek (known earlier as the Marshall House), and another Firehole name which did not survive was Firehole Pool, a hot lake east of Firehole Lake named by thermal researchers in 1935. More recently, there was a proposal to relocate the visitor facilities for the Old Faithful area in the Lower Geyser Basin as a "Firehole Village," in order to lessen the impact of visitor use of the limited area which contains most of the great geysers; but that idea was dropped early in the 1960s as the Park's Mission–66 Program (intended to rejuvenate all facilities by 1966) fizzled out far short of its goal.

## BRIDGER

While not now a Park place name in the fullest sense of the term, this word was formerly associated with the Park and has been retained in the name of a lake just outside the south boundary.

Bridger's River and Bridger's Lake appeared on DeSmet's 1851 map, and, since they are not labeled on the Bridger-DeSmet map which was the base for his compilation,[20] it seems probable DeSmet was the author, not Jim Bridger. This is entirely in keeping with Bridger's character; he was not egocentric, even in his tale-telling. It is more likely that DeSmet put Bridger's name on his compiled map by way of a "thank you!" for substantial help, or as a tribute to the mountain man's geographical knowledge.

Bridger's River found no advocates. In 1869 it appeared on a General Land Office map as "Main Fork,"[21] and in 1871 the Hayden Survey used Upper Yellowstone River, but that was dropped by the

Bridger Lake. Just outside the boundary in the southeast corner of the Park, it appeared on the manuscript map Father P. J. DeSmet prepared in 1851 from trapper Jim Bridger's information. NPS photo.

United States Geological Survey, which does not recognize that portion of the Yellowstone River above Yellowstone Lake as a separate stream.

Bridger's Lake fared better as a place name for that beautiful body of water that lies just outside Yellowstone Park, but it appears to have been a close thing. Dr. Ferdinand V. Hayden, who visited it in 1871 with a party, described the lake as "not more than one or two hundred yards in width," adding, "The lake which has been placed on the maps as Bridger's Lake has no real existence."[22] This is such a falsehood when applied to a lake over a mile in length and a half mile wide at its greatest that Hayden can be suspected of not wanting the lake to exist. Regardless, he did show Bridger's Lake on the map accompanying his report, and the United States Geological Survey subsequently (1896) turned that into the present form — Bridger Lake.

## ELEPHANT BACK

This remarkably descriptive place name from the era of the fur trappers was first applied only to a mountain near the outlet of Yellowstone Lake, but later it appeared on a formation at Mammoth Hot Springs.

Elephant Back Mountain is a place name which appeared on Captain William F. Raynolds's map of 1859–1860,[23] where it is one of a half-dozen landmarks in the conspicuously empty Yellowstone region the Raynolds Expedition passed near but not through. He was dependent upon Jim Bridger's memory for the name of that mountain which resembles the profile of an elephant's head and back, as viewed from Yellowstone Lake or its shore at Gull Point.

An error in delineating the course of the Yellowstone River north of Elephant Back Mountain led at least two Army officers to confuse Bridger's landmark with the Washburn Range. Raynolds showed the Yellowstone River beginning its prominent westward sweep immediately behind Elephant Back Mountain (that really happens about 25 miles down river, behind the Washburn Range), and that unfortunate mistake explains why both Lieut. Doane (1870) and Capt. Barlow (1871) shifted the name.[24]

Also in 1871, Ferdinand V. Hayden correctly located the "Elephant's Back" on the map of his party's first season of work in the Yellowstone region, and the United States Geological Survey confirmed the location in 1885 but spelled the name without the possessive. A final change, in 1930, completed the history of this place name when the United States Board on Geographic Names decided that the usage should be Elephant Back Mountain.

The foregoing does not include several variants of the Elephant Back name. At least one stagecoach driver called the "Knotted Forest" section of the old West Thumb-to-Lake road the "Elephant's Hump,"[25] which is an interesting

Elephant Back Mountain (8,730 ft.). Capt. William F. Raynolds was able to map this feature in 1860 from information given him by Jim Bridger. The view is from Gull Point.

twist. Another variant — Elephant Back Terrace, an early name for what is now the White Elephant Back at Mammoth Hot Springs — is unrelated to the fur trade era.

Raynold's map is the likely origin of another troublesome place name. It shows a "Sulpher Mt." on the northern verge of the Yellowstone region, and it is placed on the north bank of a stream which enters the Yellowstone just as that river resumes its northward course. Jim Bridger was the source of Raynolds' information, and, since he was also responsible for Father DeSmet's notation, "Sulphur Mountain," entered on the north bank of a "Gardeners Riv." on the 1851 map, it is obvious that the present Mammoth Hot Springs was meant.

But Raynold's cartography, being a product of hearsay, was badly warped, with an oversized Yellowstone Lake occupying too much of the plateau, and thus crowding "Elephants Back Mt." and "Falls of the Yellowstone" too far north and adjacent to the "Sulpher Mt.." The distortion should have been obvious, yet it caused such capable officers as Lieutenant Doane and General Sherman to confuse Raynold's "Elephants Back Mt." with the Washburn Range. It also led the Chief Park Naturalist, Dr. C. Max Bauer, to identify "Sulpher Mt." with the Crater Hills. He was able to get the United States Geographic Board to rename the Crater Hills — a legitimate Washburn Party place name of 1870 — as Sulphur Mountain (see *Decisions*, May 7, 1930: p. 7). The 1930 and 1954 editions of the Park-wide topographic map used only Sulphur Mountain for what had previously always been the Crater Hills.

The 1959 revision of the topographic map (*Canyon Village, Wyo.*, quadrangle) restored the Crater Hills name, but also retained Sulphur Mountain, so that the map has since presented the silly picture of a Sulphur Mountain on top of the Crater Hills. It is to be hoped that the inappropriate Sulphur Mountain name will some day be dropped as the Park already has enough "sulphurous" names.

## TWO OCEAN/TWO WATER

The origin of these place names lay in a phenomenon outside the present Yellowstone Park, but known among the trappers at least as early as 1836. It has since been applied to some features within the Park.

Two Ocean River is the original form, as we have it on the DeSmet map of 1851 (compiled from Jim Bridger's crude sketch of the Yellowstone region).[26] However, trapper Osborne Russell, who paused at Two Ocean Pass on August 9, 1836, described the stream and its locale very well, even though he did not use the name. He wrote in his Journal:

> we came to a smooth prarie about 2 Mls long and half a Ml wide lying east and west surrounded by pines. On the South side about midway of the prarie stands a high snowy peak from whence issues a Stream of water which after entering the plain it divides equally one half running West and the other East thus bidding adieu to each other one bound for the Pacific and the other for the Atlantic ocean. Here a trout of 12 inches in length may cross the mountains in safety. Poets have sung of the "meeting of waters" and fish climbing cataracts but the "parting of waters and fish crossing mountains" I believe remains unsung as yet by all except the solitary Trapper who sits under the shade of a spreading pine whistling blank-verse and beating time to the tune with a whip on his trap sack whilst musing on the parting advice of these waters.[27]

That description is essentially correct, though modern mapping shows two streams, a North Two Ocean and a South Two Ocean, descending from mountains flanking opposite sides of the pass to create a marshy prairie out of which flow Atlantic Creek (toward Yellowstone River) and Pacific Creek (toward Jackson Hole and Snake River). The creeks do interconnect in the marsh.

Although Russell did not name the pass, it was known as the "Yellowstone Pass" by the trappers because it served as a southern gateway into the Yellowstone region.[28] Captain Raynolds' map (1860) shows a "Pass Creek or Two Ocean Water" at that point. Again, Jim Bridger was certainly the source. It did not become Two Ocean Pass until surveyed by the United States Geological Survey in 1885.

But the name — Two Ocean Pass — was used earlier by the Hayden Survey, which applied it to the low pass between Grouse and Outlet Creeks in Yellowstone Park.[29] The name did not stick and that pass is unnamed on current maps.

Another place name which has not persisted was "Two Ocean Pond," an ephemeral body of water between Upper Geyser Basin and West Thumb which appeared on Superintendent P. W. Norris's 1879 map and was probably named by him.[30] This may have been an attempt to give substance to Lord Dunraven's "Two-water Lake" (1874), another hydrological phantom of that area. However that may be, there is only one other reference to this elusive pond during the early period. A learned Englishman travelling in the Park in 1883, said about it: "This I should much like to have seen, for a lake with *two outlets* is, I believe, an impossible *lusus naturae*, existing in imagination only."[31] His skepticism was justified, for several generations of Army scouts and Park rangers searched fruitlessly for "Lost Lake," as it appeared on official maps until aerial mapping proved beyond doubt there was no such body of water. It must be added that Historian Chittenden preferred to identify that Lost Lake with a pond mentioned by Dr. Hayden in his 1871 report: "We camped at night on the shore of a lake which seemed to have no outlet. It is simply a depression which receives the drainage of the surrounding hills." Perhaps the answer to this puzzle lies in the transient nature of a melt-water pond which has not since been seen at just the right stage.

Two approved variants of the Two Ocean place name appear on current maps of the Park. Two Ocean Plateau, the high land on the west of the Yellowstone River which extends northward from Two Ocean Pass into Yellowstone Park, was named by the United States Geological Survey in 1885, and Two Ocean Plateau Trail is an administrative name for a route up Passage Creek.

## HART/HEART

Which form is right will likely never be known; that is, whether or not this place name immortalizes a trapper, or an anatomical resemblance. But the latter does have the edge through common use.

Historian Chittenden says that the beautiful lake lying at the foot of Mount Sheridan and known as Heart Lake was "named prior to 1870 for an old hunter by the name of Hart Hunney who in early times plied his trade in this vicinity."[32] He cites Ferdinand V. Hayden as his authority, and adds that Hart was one of Bonneville's trappers killed by Crow Indians in 1852. Whatever truth there might be in either statement appears beyond verification as Hart's name has not been found in extant records of the fur trade.

Why Heart Lake was given its present name is no mystery. Captain John Barlow, of the Barlow-Heap Expedition of 1871, named it on August 10, noting that it was so-called "from its shape."[33] Although Chittenden has protested that there is no similarity between the outline of the lake and the form of a heart, the usage has never been seriously questioned.

Related variations of this name include Heart River (1878), the outlet stream connecting Heart Lake with Snake River; Heart Lake Geyser Basin (1878), a thermal area scattered along several miles of Witch Creek immediately above the Lake, and several administrative usages — Heart Lake Ranger Station (1928), later Heart Lake Patrol Cabin (1959), and Heart Lake trail (1956).

## COLTER/COULTER

These place names are drawn from the surnames of two different men, though they are frequently confused. One was a trapper and the white discoverer of the Yellowstone wonders, while the other was an early scientist.

Colter Peak is the only place name within Yellowstone Park which immortalizes John Colter, that hardy former member of the Lewis and Clark Expedition turned fur trapper, who passed through the Yellowstone wilderness on a solitary trek in the winter of 1807–1808 and became the first white man to see some of its thermal features. The peak was so-named by the United States Geological Survey in 1886, though Mount Langford would have been a more appropriate name because it is the one he and Lieutenant Doane climbed on September 7, 1870, in order to view the country the Washburn party was passing through.

Langford and Doane returned to the expedition's camp late at night with a sketch of Yellowstone Lake and the country to its east and south; an accomplishment which is noted in Langford's *Diary* in these words:

> General Washburn and Hauser, as well as myself, have thus far made outlines of the lake shore as best we could from points on a level with the lake, but these have been unsatisfactory and have lacked completeness, and Washburn and Hauser have both expressed their satisfaction with the sketch of the lake shore I made to-day from the top of the mountain; and Washburn has just told me that Lieutenant Doane has suggested that, as I was the first to reach the summit of the mountain, the peak should be named for me. I should be gratified if this is done.[34]

A footnote to the above citation shows what happened. As Langford states, Washburn, in his official capacity as the Surveyor-General of Montana Territory, did submit a revised map to the Interior

Department. But that rough sketch placed Mount Langford somewhat too far to the north (in this it was faithful to Langford's sketch made from the mountain top).[35] That defect was picked up by the Hayden Survey, which placed the name on the wrong peak. Langford blamed Dr. Hayden for that, complaining that "on the map which he issued in connection with his 1871 report, the name 'Mount Langford' was given to another mountain far to the northeast. . . . I think that Dr. Hayden must have been aware at that time that this mountain bore my name."[36] The error was not corrected and that left an unnamed peak for the United States Geological Survey to put John Colter's name on in 1885.

Colter Peak (10,683 ft.). Named for trapper John Colter, who appears to have been the first white man to enter what is now Yellowstone National Park while on a winter journey in 1807–1808.

Colter's name has been improperly associated with the Park in such literary usages as identify "Colter's Hell" with the thermal features on the Park plateau. The locality the trappers knew by that name was near the present town of Cody, Wyoming. Odoriferous hot springs at the base of Cedar Mountain gave the Shoshone River its original name of the "Stinkingwater." (The DeMaris Springs on the outskirts of Cody are a remnant of this thermal area, much of which is now under the waters impounded behind Buffalo Bill Dam).

Historian Merrill Mattes has this to say about the "Colter's Hell" place name: "There *was* a Colter's Hell in the fur trappers lexicon which referred specifically to an ancient thermal area bordering the Shoshone River just west of present Cody, Wyoming. The term was *never* applied historically to the thermal zone within Yellowstone Park itself. It was Hiram M. Chittenden, the esteemed engineer and historian, who first suggested this usage in 1895 with the original edition of his book, *Yellowstone National Park*."[37]

Concerning the use of "Colter's Hell" as a literary cognomen for the Park, Mattes adds: "No item of Yellowstone history is more widely believed, more universally beloved, and more transparently incorrect."

Another form of this place name — Coulter — is also often mistakenly cited as evidence of John Colter's presence in Yellowstone Park. Coulter Creek, a stream which runs into Snake River just within the south boundary of the Park, was named by the United States Geological Survey in 1885 for John Merle Coulter, a botanist with the Hayden Survey in 1872. Discovery on the bank of the stream, in 1889, of a large pine blazed with a large "X," and below that the initials "JC," led hunting guides to presume the tree had been marked by John Colter during his winter journey across the Yellowstone plateau in 1807–1808.[38]

But the classroom notebook of one of Dr. Coulter's students at the University of Chicago, where he taught botany in later years, provides a logical explanation for the blaze. The student wrote: "Dr. Coulter was fishing one day on the bank of a stream when he felt a slap on the shoulder and turned expecting to see one of his companions, but there was a large, black bear. Coulter plunged in and swam across the stream, then looking back saw that the bear had not followed, but was back there grinning at him. The party called this stream Coulter Creek, a name it still bears."[39] Thus, the "X" appears to mark the spot where J. C. — John Merle Coulter — took his swim!

## SULPHUR/SULPHER

Several of the Park's place names involving these words are of trapper origin, but not all.

"White Sulphur Banks" appeared on the Washington Hood map of 1839, presumably from fur trade informants since Merrill Mattes credits "William Sublette and others."[40] The indicated feature is in the right relationship to "Gardner's Fork" to represent the Mammoth Hot Spring.

The DeSmet map (1851) shows a "Sulphur Mountain," also in the right place to represent Mammoth Hot Springs; as is the "Sulpher Mtn." on Raynolds's map (1860). Since both maps were compiled with Jim Bridger's help in filling in details of an area not visited by the authors, they must represent fur trade knowledge.

There is no relation between the foregoing place names and similar names — Sulphur Mountain in Hayden Valley, and Sulphur Hills near Fishing Bridge — on current maps of the Park.

## MISCELLANY

The place names in this group are trapper related only in the sense that they are, with perhaps one exception, latter-day references to the fur trade.

Alum Creek is a place name which appears to have had its genesis with the prospectors — applied to a stream draining the Mirror Plateau, possibly Broad Creek — though it came into its present usage with the explorers (see Chapters 3 and 4). However, its popular identification with a Jim Bridger "story" concerning the marvelous ability of waters of a Hayden Valley stream to shrink almost anything has given this name an undeserved relationship to the fur trade. The story, which appears in several forms, is *not* a confirmed part of Jim's repertoire, but only a modern literary accretion. Actually, it made its debut in 1903, in the "revised and enlarged" fourth edition of Chittenden's book *The Yellowstone National Park*, where the story goes this way (pp. 49–50):

> The origin of the name of Alum Creek, a tributary of the Yellowstone, was due to an accidental discovery by Bridger. One day he forded the creek and rode out several miles and back. He noticed that the return journey was only a small fraction of the distance going, and that his horse's feet had shrunk to mere points which sank into the solid ground, so that the animal could scarcely hobble along. Seeking the cause he found it to be in the astringent quality of the water, which was saturated with alum to such an extent that it had the power to pucker distance itself.

By 1960, the tale had been embroidered with this unlikely soliloquy:

> "We was ridin' east o' the river along the side o' a creek. 'Twas boggy an' goin' was slow. When the cañon narrowed we guide our hosses into the stream. 'Twasn't no time till the hosses' feet shrunk to pin points, an', by tarnation, we went twenty miles in a jiffy. Them thar waters was so strong o' alum as to pucker distance itself!"[41]

Cache Lake, located at the head of Reese Creek is a place name of uncertain origin. The Hayden Survey had previously shown Reese Creek as "Cache Creek,"[42] a place name which suggests the presence of a "cache" — the type of underground structure most commonly employed by trappers to safeguard their supplies and furs, though occasionally also used by prospectors. The lake was so-named by the United States Geological Survey in 1883, at the same time that the stream name was changed.

Potts Hot Spring Basin, a place name applied administratively in 1956 to a small area of hot springs 1.2 miles north of the old West Thumb road junction, honors Daniel T. Potts, an American trapper whose letter to home folks became the earliest published descripton of the Park's great lake and Yellowstone's geyser activity. Spelling and punctuation are shown here as in the MS letter:

> the Yellow-stone has a large fresh water lake near its head on the verry top of the Mountain which is about one hundrid by fourty miles in diameter and as clear as crystal on the south borders of this lake is a number of hot and boiling springs some of water and others of most beautiful fine clay and resembles that of a mush pot and throws its particles to the immense height of from twenty to thirty feet in height The clay is white and of a pink and water appear fathomless as it appears to be entirely hollow under neath. There is also a number of places where the pure suphor is sent forth in abundance one of our men Visited one of those wilst taking his recreation there at an instan the earth began a tremendious trembling and he with dificulty made his escape when an explosion took place resembling that of thunder. During our stay in that quarter I heard it every day.[43]

Trappers Creek is a place name of a generic type intended to immortalize the men of the fur trade. The name which the United States Geological Survey gave this stream in 1885 is quite appropriate, located as it is immediately southeast of Colter Peak — named for the trapper who was Yellowstone's white discoverer.

Brief mention of a few trapper usages which did not survive the fur trade era will do to close this section.

Trapper Osborne Russell, in the course of his several "hunts" in the Yellowstone region, named some Park features in the diary he kept. These included "Secluded Valley" (1836), that part of the Lamar

Valley occupied by the Buffalo Ranch in later years; "Hour Spring," a geyser in the Shoshone Geyser Basin (1839), and "Boiling Lake," now Grand Prismatic Spring in Midway Geyser Basin (1839). Although these names are known only from Russell's *Journal*,[44] they were probably used by other trappers.

Two more place names which can be attributed to Jim Bridger are shown on DeSmet's map (1851). They are the "Little Falls," a contradistinction separating Tower Fall from the greater Yellowstone Falls, and "Beaver Creek," now Lamar River. These features are not accurately related on the DeSmet compilation, but reference to the more accurate hydrography of the Bridger-DeSmet map leaves no doubt as to the intent.

Another place name of trapper origin — "Piney Woods" — was recorded by Warren Angus Ferris. It was descriptive of the forest on the Madison Plateau, which Ferris crossed on May 19, 1834, during a visit to the geyser basins on the Firehole River.[45] This usage may have been uniquely his.

Lamar River. Looking eastward up the beautiful "Secluded Valley" described by trapper Osborne Russell in 1836. The present name honors L.Q.C. Lamar, a former secretary of Interior.

## NOTES

1. William Clark, *Map of Part of the Continent of North America*, Coe No. 303-iv, Beinecke Library, Yale University, or the Samuel Lewis engraving from that source (1814).
2. Hiram M. Chittenden, *The American Fur Trade of the Far West* (New York: Francis P. Harper, 1902), II, pp. 941–945. The original accounts are in the Chouteau-Maffitt Collection, Missouri Historical Society, St. Louis.
3. Osborne Russell, *Journal of a Trapper*, ed. by Aubrey L. Haines (Portland: Oregon Historical Society, 1955), pp. 28, 46 & 101.
4. Ibid., p. 46; see the Hood map (1839), and DeSmet's maps (1851), for the Gardner's Fork usage, and the General Land Office maps (1869), deLacy map (1870), and Barlow map (1872), for the Gardner's River usage.
5. Hiram M. Chittenden, *The Yellowstone National Park*, (Cincinnati: Robert Clarke Co., 1895), p. 317.
6. Chittenden, *Yellowstone* (1895), pp. 317–318.
7. Nathaniel P. Langford, *Diary of the Washburn Expedition to the Yellowstone and Firehole Rivers in the Year 1870* (St. Paul, MN: F. J. Haynes Co., 1905), p.viii.
8. The cemetery, a little less than a mile west of town on the county road, was established in the 1890s, its first occupant being an itinerant tinsmith crushed to death while removing the wheel from a freight wagon.
9. Jack E. Haynes, Memorandum to the files, Jan. 8, 1953, concerning information on territorial post offices received from Jefferson Jones, Dec. 1952.
10. S. P. Panton, "Early Days in Yellowstone recalled by N.P. Surveyor," *Wolf Point* (MN) *News*, Nov. 23, 1939.
11. Warren A. Ferris, *Life in the Rocky Mountains, 1830–1835*, P.C. Phillips, ed.(Denver: Old West Publishing Co., 1940), pp. 85–86. Ranger Richard Halman, of the West Yellowstone District, Gallatin National Forest, stated on February 8, 1967, that the results of forest inventory studies indicated an even-age stand in the Madison Basin, with an origin at about A.D. 1817. He considered this evidence of an extensive burn.
12. Russell, *Journal*, pp. 29 & 100; see Note 3 for citation.
13. William F. Raynolds, "The Report of Brevet Brigadier General W. F. Raynolds on the Exploration of the Yellowstone and the Country Drained by That River," *United States, 40th Cong., 1st Sess. Senate. Ex. Doc. No. 77*, ordered to be printed July 17, 1868, p. 99 [Entry for June 25, 1860]. The map accompanying Raynolds's report locates the Burnt Hole correctly with reference to the foregoing information. The original is in National Archives, RG–77; Office of the Chief of Engineers, Q 106–1.
14. Washington Hood, "Map exhibiting the practicable passes of the Rocky Mountains," prepared by Washington Hood, Captain, Corps of Topographical Engineers, 1839 [information supplied by William Sublette and others]. Original in the National Archives, RG–77, Office of the Chief of Engineers, US 110.
15. The originals of these two maps are in the Missouri Province Archives (Jesuits), at St. Louis University, Arch. IX: DeSmetiana. C–8: Atlas. Bridger-DeSmet, 1851, is No. 10, and DeSmet, 1851, is No. 5.

16. The original manuscript map is in the collection of the Montana Historical Society, Helena (map case 2, drawer 2). Printings by Jul. Hutawa, St. Louis; Rae Smith, New York, and Fridenwald, Baltimore, are known. Also, Hutawa's version appeared at reduced size in Granville Stuart's *Montana as it Is* (1865).
17. See one of the several printings mentioned in the previous note.
18. Gustavus C. Doane, "Map of the Route of the Yellowstone Expedition Escort Commanded by Lieut. G. C. Doane, U.S.A., Sept. 1870," National Archives, RG–77, Office of the Chief of Engineers, Q 329 — No. 30.
19. Gustavus C. Doane, "Official Report of the Washburn-Langford-Doane Expedition into the Upper Yellowstone in 1870," 78 pp. A handwritten copy made at Fort Ellis, MT, Dec. 15, 1870. Montana State University Library MS 492, p. 53.
20. DeSmet, 1851, No. 5; see Note 15 for citation.
21. General Land Office, Manuscript map prepared by H. D. Washburn, 1869; National Archives, RG 49, GLO (OMF) — Montana 3.
22. Ferdinand V. Hayden, *Preliminary Report of the United States Geological Survey of Montana and Portions of Adjacent Territories, Being a Fifth Annual Report of Progress* (Washington: Government Printing Office, 1872), pp. 132–133.
23. Raynolds, "Map"; see Note 13 for citation.
24. Doane, "Official Report," p. 16; see Note 19 for citation. John W. Barlow and David P. Heap, *Report of a Reconnoissance of the Basin of the Upper Yellowstone in 1871*, (Washington: Government Printing Office, 1872), p. 13.
25. Yellowstone C. Samuell. "Heavers, Hashers, Skinners & Dudes," *Montana Magazine of History* (Jan. 1980), p. 52.
26. See Note 15, map No. 10.
27. Russel, *Journal*, pp. 43–44; see Note 3 for citation.
28. Hood, "Map"; see Note 14 for citation.
29. Hayden, *Fifth Report*, p. 133; see Note 22 for citation.
30. Philetus W. Norris, *Report Upon the Yellowstone National Park to the Secretary of the Interior. . . for the Year 1879* (Washington: Government Printing Office, 1880), p. 29.
31. Charles T. Whitmell, "The American Wonderland, The Yellowstone National Park," *Report and Transactions of the Cardiff Naturalist's Society*, XVII [1885](Cardiff, Wales: South Wales Printing Works, 1886), pp. 77–106.
32. Hiram M. Chittenden, *The Yellowstone National Park*, ed. by Richard A. Bartlett (Norman: University of Oklahoma Press, 1964), p. 174.
33. Barlow and Heap, op. cit., p. 33. The manuscript compiled following the expedition identifies the lake; see, National Archives, RG–77, Headquarters map file, Q 181, Office of the Chief of Engineers.
34. Langford, *Diary*, p. 65; see Note 7 for citation.
35. Henry D. Washburn, "Route of the Washburn Party, 1870, being a tracing of map made to accompany Washburn's report to Dept. of Interior," National Archives, RG–49 (OMF), Records of the Gen. Land Office, Mont. 5 Cf. *ante*, p. 64.
36. Langford, loc. cit., p. 66.
37. Merrill J. Mattes, *Colter's Hell & Jackson's Hole* (Yellowstone Library and Museum Association, 1962), p. 19.
38. Stallo Vinton, *John Colter, Discoverer of Yellowstone Park* (New York: E. Eberstadt, 1926), pp. 61–62.

39. Dorothea G. Doubt (Chicago) to "The Secretary of the Museum" (Mammoth Hot Springs), Sept. 7, 1933.
40. Mattes, op. cit., p. 51.
41. Merrill D. Beal, *The Story of Man in Yellowstone* (Yellowstone Park, Wyo.: Yellowstone Library and Museum Assoc., 1960), p. 113.
42. Ferdinand V. Hayden, [map] "Yellowstone National Park," 1871. National Archives, RG–57, Hayden Survey, YNP.
43. Daniel T. Potts, Letter to Robert T. Potts, July 8, 1827, *Philadelphia Gazette and Daily Advertizer*, Sept. 27, 1827. The original letter is held by the Yellowstone Park Reference Library.
44. Russell, *Journal*, pp. 26, 46, 62, 66, 98–100; see Note 3 for citation.
45. Ferris, *Life in the Rocky Mountains*, p. 258; see Note 11 for citation.

# 3

# SOME PROSPECTOR ADD-ONS

*Wonderland*
*Pelican*
*Crevice*
*Slough*
*Turkey Pen*
*Miscellany*
*Cache*
*Snowy/Snow*
*Hell Roaring/Hellroaring*
*Soda Butte*
*DeLacy*

The Montana gold rush of 1862 was the magnet that attracted a large population of prospectors and miners to the northern Rocky Mountains. They came from the earlier diggings in Idaho and Colorado, and from the "States" where the Civil War had gone on long enough to disenchant many men, and this rough horde soon overtaxed the original strike at Bannack. From there prospecting parties spread out to discover other placers and establish other settlements, a process that continued to the end of the 1860s. The Yellowstone plateau was not exempt from this thorough ransacking, and the passage of prospecting parties through its wilderness left some place names reminiscent of their experiences.

Of the many parties of prospectors afield in 1863, one is important as a precursor of more successful invasions of the Yellowstone headwaters. This party banded together in late summer to examine the upper reaches of the Snake River under the leadership of Walter W. deLacy, a mild-mannered civil engineer who had little in common with the "forty thieves," as he called his comrades. Entering what is now Yellowstone Park by its back door — that is, from Jackson Hole — this party and a splinter group passed through the Shoshone and Firehole geyser basins without finding gold. But they did return to Bannack with valuable geographical information that deLacy incorporated into his *Map of the Territory of Montana*, published in 1865.

Prospecting in 1864 resulted in three incursions into the Yellowstone country. One led by George A. Huston went up the Yellowstone River from the mouth of Emigrant Gulch (then called Curry's

Gulch after the man who made the strike in 1863), prospecting the Lamar and Clarks Fork drainages. Another, the Phelps–Davis party, entered what is now the Park by way of the old trapper's route over Two Ocean Pass. These men followed the east shore of Yellowstone Lake to its outlet, then split up to examine both rims of the Grand Canyon before passing downriver to the settlement at Emigrant Gulch — the short-lived Yellowstone City.

The prospectors who had prowled the Yellowstone country that summer made their way to Alder Gulch, passing en route six cabins on a stream in the Gallatin Valley (the beginnings of an agricultural settlement that would soon be known as Bozeman). A rich strike made the previous fall had already attracted a large number of prospectors to Alder Gulch, and that influx led to organization of Virginia City — another of the rude towns that served the mining frontier as staging points for further prospecting expeditions.

George Huston immediately organized another party that ascended the Madison River and the Firehole as far as the geyser basins, where they intended to pass the winter; but the cold, vapor-laden air of late fall impressed them as being suffocatingly noxious, and they retreated from that inhospitable place. The only positive result of that expedition — the pioneering of a new route into the Yellowstone country — was offset by the persistent myth of a "death valley."

In 1865, prospecting on the Yellowstone plateau seems to have been largely suspended in favor of new strikes at Last Chance Gulch (Helena) and Confederate Gulch (Diamond City). Available records indicate only one visit to the Upper Geyser Basin by prospector George Harvey Bacon and a party of Indians.

George Huston returned in 1866 with five companions, again entering the area by way of the Madison River. This time they did not linger among the geysers but proceeded to Hayden Valley by way of Nez Perce Creek, and then circled Yellowstone Lake counterclockwise before descending from its outlet to the Yellowstone Falls. From the falls they crossed Mirror Plateau to Lamar River and continued down that stream and the Yellowstone to Emigrant Gulch (where a new town — Chico — had replaced the ephemeral Yellowstone City). This extensive tour seems to have convinced the prospectors that the interior of what is now the Park did not contain the mineral wealth they were seeking.

But the year ended with great expectations for the mountainous northern and eastern rims of the Yellowstone plateau. In that region,

"Uncle" Joe Brown and three others extracted gold dust and nuggets worth $8,000 from a river bar at the mouth of Bear Creek (just outside the present Park).

The year 1867 was one of Indian hostility on Montana's eastern approaches. This condition was generated mainly by settlers using Bozeman's "road," that wagon route intended to give access to Montana Territory from the Oregon Trail, and by the forts built to protect traffic along it. The effect was to turn the attention of prospectors back to the Absaroka Range (the "Snowy Mountains" in that day), where they were safer from the hostile Sioux. This resulted in several more visits of prospecting parties to what is now the Park.

A company of Montana's territorial militia made a reconnaissance trip as far up the Yellowstone River as the Mammoth Hot Springs without encountering Indians, and that tempted a party of prospectors to go farther. They found gold at Crevice Creek, continued up the river and its east fork (Lamar River), and then crossed Mirror Plateau to Pelican Creek and Hayden Valley. From there they followed the Nez Perce Creek–Firehole–Madison River route to Virginia City.

Another party of Montana prospectors, including A. Bart Henderson, originated at Deer Lodge and took the Utah stage road as far as the Snake River Plain in Idaho; from there they went up the Snake River into Jackson Hole and crossed Two Ocean Pass into what is now the Park. Their route through the area was by the east shore of Yellowstone Lake and down the river to the Montana settlements.

In a way, "settlement" could be considered to have crept to the very door of the present Park, for George Huston took up residence opposite Bear Creek that fall, and there were three miners working placers at the mouth of Crevice Creek. And on the west, or Idaho side, Gilman Sawtell had located a "ranch" at Henrys Lake.

Another ranch, this one in the Yellowstone Valley across from Emigrant Gulch, was established in 1868. Two brothers, Frederick and Phillip Bottler, held "bachelor hall" there when not off on hunting and prospecting ventures — mostly into the Yellowstone mountains. They were with a party that went as far as Sunlight Basin in 1869, and there found the despoiled camp and murdered remains of two of the Crevice Creek miners — Crandall and Dougherty — who had moved over too close to Indian country.

A. Bart Henderson was back in 1870 with a party that discovered the mines on Clarks Fork (the present Cooke City area). That put an end to prospecting in the Yellowstone interior by diverting the attention of miners to the mountains north and northeast of the present Park.

The brief era of the prospector was a time of remarkably far-ranging, intensive activity, and it is amazing that so few place names remain from what must have been a very detailed geographical knowledge.

## WONDERLAND

Of all the place names known to have a prospector origin, none is quite as important as "Wonderland" — used as a synonym for the Yellowstone region and later the Park established there. Implicit in the word is something of the feeling of awe Anglo-Saxon people felt for the unusual and mysterious.

That such a rough-and-tumble prospector as A. Bart Henderson should be the first to refer to the Yellowstone region by a term popularized through a children's story is something to wonder about in itself, for "Wonderland" was as yet a new usage.[1] His diary entry for July 22, 1871, states: "Clear & warm. Left camp early, arrived at Mammoth Hot Springs at noon. Here I found Dr. Hayden & Col. Barlow, on their way to Wonderland. Here I remained until Monday, bathing & looking around." The entry for the 24th is the last for the year and again refers to Wonderland: "Left camp at 9 o'clock & followed down the river. arrived at Bottlers Ranch. Here I remained a few days, resting and viewing out a road which I located on the 12 day of Aug. 1871. It is to run from Bozeman to the Yellowstone Lake, by the Mammoth Hot Springs, built for the benefit of the travel to & from Wonderland, & to be a toll road. I soon commenced work on the same, where I remained till Christmas."[2]

The first published use of Wonderland in reference to the Yellowstone region appeared in the Deer Lodge, Montana, newspaper in September, and it appeared in the eastern press in October. Both uses were in titles.[3] Thus, the area had a sobriquet months before it became a Park, which tends to compensate for the fact that it remained without an official name for months afterward.

It must be added here that the little girl for whom Dodgson (Lewis Carroll was the author's pen name) created his "Wonderland" tale was able to visit Yellowstone Park as a grown-up, and she seemed as thrilled as if she had really gotten into that peculiar storybook place through the rabbit hole.

## CACHE

All the place names involving this word are related to prospector usage except Cache Lake at the head of Reese Creek.

The earliest — Cache Creek — received its name from an unfortunate incident involving a party of prospectors in the spring of 1864. There are two versions of this event, and they are in reasonable agreement. The earlier and better of the two was set down by Superintendent P. W. Norris, who knew several of the participants personally. His account is used here:

> In the spring of 1864, H. W. Wayant, now a leading citizen of Silver City, Idaho, William Hamilton, and other prospectors, to the number of forty men, with saddle horses, pack train, and outfit, ascended the east side of the Yellowstone from the Gate of the Mountains to Emigrant, Bear and Crevice Gulches, forks of the Yellowstone, East Fork, and Soda Butte; thence over the western foothills of Mount Norris to the bluffs upon the south side of Cache Creek, where their horses were all stolen by some unknown Indians, but their only two donkeys would not stampede, and remained with them. Here the party broke up; Wayant, Harrison, and ten others, with one jack, and what he and the men could carry, ascended Cache Creek to Crandall Creek, Clark's Fork, Heart Mountain, thence by way of Index Peak and the Soda Butte returned to the cache made by the other party of what they could not carry, aided by their donkey, from where set afoot, and hence called Cache Creek.[4]

Related place names are Cache Mountain, named by the Geological Survey (1885) because it is nearly enclosed by the headwaters of Cache Creek, and Cache Creek Patrol Cabin, an administrative usage for a structure midway up the stream.

## PELICAN

Of the many place names involving this word, all but one are related to prospector usage.

Pelican Creek, the original usage, also developed out of an incident in 1864. One of the men who passed through the southwest corner of what is now the Park with deLacy in 1863 came back the following year with a small party that entered the area by way of Two

Ocean Pass. He was John C. Davis, and he told a Louisville, Kentucky, reporter this story twenty years later:

> We camped on this creek, and I noticed several large birds which appeared to be wild geese. I shot one, which managed to fly out some distance in the lake before it fell. I swam out after it, and became very much exhausted before I reached it. It looked as if it might be good to eat so I skinned it, and then the boys concluded it would hardly do. I hung the Pelican — for that was what it was — on a tree, and it was found afterward by Miller, who came by with his party.[5]

The stream that debouched into Yellowstone Lake nearby became known among the prospectors and miners as Pelican Creek, and the Hayden Survey picked up that name from some of the men who were employed as packers and hunters in 1871. It appeared on the map of the 1871 field work,[6] and has been used on all maps of the Geological Survey in that form.

Place names that the Hayden Survey soon associated with the name of the creek are Pelican Hill and Pelican Creek Mud Volcano,

Pelican Creek. A pelican killed by the John Davis party of prospectors in 1864 where this marshy stream enters Yellowstone Lake was left hanging on a tree, and thus provided the stream name.

both in 1878. Pelican Hill, which P. W. Norris labeled in the possessive form on his 1879 map, was renamed Pelican Cone by the Geological Survey (1904). Pelican Creek Mud Volcano is a feature that cannot be identified on present maps.

The Geological Survey put Pelican Springs on the map in 1885, and Historian Chittenden made use of the term "Pelican Valley" as early as 1915;[7] however, it did not appear on official maps of the Park until 1959. Several other administrative usages made their appearance at that time: Pelican Creek Campground, Pelican Creek Trail, and Pelican Cone Trail.

Another place name — Pelican Roost — is in no way related to the Pelican Creek name. It was used by the Hayden Survey to designate a bird rock in Yellowstone Lake because of the Pelicans seen there.

## SNOWY/SNOW

These are alternative forms of a place name designating what is now the Absaroka Range.

The editor of the first newspaper in Montana Territory, James Hamilton Mills (himself an ex-prospector), wrote: "The Scenery of the Yosemite Valley, as described by Bowles in his new book, 'Across the Continent,' though very grand and peculiar, is not more remarkable than the scenery at the passage of the Yellowstone through the Snowy Range, one hundred miles northeast of this city."[8] That the editor's usage — Snowy Range — was not just a creation peculiar to himself is shown by the appearance of its alternative form, Snow Mountains, on the General Land Office map in 1869.[9] The replacement of this name by other usages and the ultimate supremacy of the present Absaroka Range are outlined in Chapter 1 under Absaroka.

## CREVICE

All the place names involving this word are related to a descriptive usage of the prospectors.

The original form of this place name — Crevice Creek — appeared on a General Land Office map in 1869.[10] However, the story behind the name was not published until 1883, when E. S. Topping produced a little book that is a potpourri of Yellowstone fact and gossip. In it he writes: "Early in the summer of 1867, Lou Anderson . . . with Hubble, Reese, Caldwell and another man, went up the river on the east side. They found gold in a crevice at the mouth of the first stream above Bear, and named it in consequence, Crevice gulch."[11]

Crevice Gulch. Mining debris left by prospectors who illegally reopened the old mining works on Crevice Creek during the hard times of the Great Depression.

That ambiguous statement lost sight of the fact that "gulch" is not equivalent to "creek"; in fact it is a small ravine, which in miner's parlance denoted a working area. While the stream was sometimes called Crevice Creek, more often the usage was Crevice Gulch, and that was encouraged by the Geological Survey through its use of the improper form on Park maps until 1955. The stream name is now officially Crevice Creek.

It needs to be added that the Crevice Gulch usage was *not* Topping's invention. The Barlow-Heap map (1871) is the earliest to show a stream so labeled, but only William Blackmore explains the true nature of the miner's usage. He wrote in his diary, "Passed crevice gulch where nugget worth 10 £ was found — placer diggings above very rich 500 H per day but cannot be worked for want of water."[12] While the Crevice Gulch usage is of some historical importance, it is no longer a proper place name.

Crevice Mountain is not really a Park place name as it is only one of several named summits on the massif rising up out of the Park from the Yellowstone River. Still, it was the source of gold found in the Crevice Gulch placer below. The name first appeared in an 1880 edition of the Bozeman newspaper. According to the editor, placer mines in Crevice Gulch were never extensive, and with the discovery of quartz veins on Crevice Mountain in 1879, mining activity in that area shifted to hard-rock claims.[13] One of the miners who took a quick profit from the mountain was "Uncle" John F. Yancey. He sold his claim and used the money to establish himself in Yellowstone Park, at Pleasant Valley on the stage road to Cooke City.

It is difficult to account for the place name of Crevice Lake. It did not appear on Park maps — official or otherwise — until 1955, when the Geological Survey placed it on the Gardiner quadrangle of the new topographic map series. Yet I personally remember that the lake — a respectable body of water — was called this by the rangers in 1938, and the name is probably much older.

The only administrative use made of the Crevice name was in designating the ranger station built in 1912 just inside the Park boundary where it passed over a shoulder of Crevice Mountain. The station, which the Army intended as a replacement for the old Knowles Station at the mouth of Crevice Gulch, is referred to as Crevice Ranger Station in the *Monthly Report of the Park Superintendent* for July 1921, and was so known until changed to Crevice Station on the 1955 revision of the Park topography (Gardiner quadrangle).

## HELL ROARING/HELLROARING

An ambiguous mapping of this prospector usage has led to confusion in both form and location.

Fortunately, the circumstances under which Hellroaring Creek was named are known, so that there should be no doubt about the correct form or the location of the stream. Yet after considerable vacillation the form has come out wrong, but the place *is* right — of which it might be said, "you win some, you lose some!"

The original form, Hell Roaring Creek, is another place name given by Lou Anderson's party of prospectors in 1867. E. S. Topping's account of the progress of that party up the Yellowstone River continues this way: "Hubble went ahead the next day for a hunt and upon his return he was asked what kind of a stream the next creek was. 'It's a hell roarer,' was his reply, and Hell Roaring is its name to this day."[14]

The name made its first formal appearance on a General Land Office map in 1869 (the earlier of the two submitted to the commissioner that year; identifiable by a spindle-shaped Yellowstone Lake), and the way in which the name was printed may have led to the later practice of spelling it as one word. On the General Land Office map

Hellroaring Creek. Named by prospectors who first saw the stream in 1867 during a period of turbulent runoff. Their impression is not characteristic of the stream.

it appeared as "Hell roaring Ck.," and, whether the separation was intended or merely inadvertent — as failure to capitalize roaring hints — the Geological Survey adopted "Hellroaring" for this and related place names in 1885, and that has been the approved form since that time.

In the matter of location the result was better. That the Washburn Expedition of 1870 had a misconception of the location of Hell Roaring Creek is evident from the fact that both General Washburn and Lieutenant Doane applied the name to what is now Sour Creek in Hayden Valley (Doane labeled it a "river," perhaps because he viewed it across the Yellowstone and the stream's broad estuary might have seemed large enough to justify that term). The 1871 General Land Office map also puts Hell Roaring Creek in Hayden Valley, which is amazing in view of the fact that other government maps of that year — Hayden's and Barlow's — had the stream back where it belonged.

Two place names related to the name of the creek are Hellroaring Mountain, a conical, rocky peak west of the stream, and Hellroaring Ranger Station on the right bank two miles above its mouth. The mountain's name was assigned by the Hayden Survey in 1871 (later condensed to Hellroaring Mountain by the Geological Survey), and the ranger station made its appearance on the Park-wide topographic map in 1928. The building was there from army days but was previously marked only as a "snowshoe cabin."

## SLOUGH

This place name has something of a chameleon character; most of the time it is apt enough, but it can occasionally be quite the opposite, as happened at the very time of its naming.

Slough Creek is the third of three names left us by Lou Anderson's party in 1867. Again, it is Topping who saved the story, which we pick up in his account following the naming of Hellroaring Creek:

> The second day after this he [Hubble] was again ahead, and the same question being asked him, he said 'Twas but a slough.' When the party came to it they found a rushing torrent, and in crossing, a pack horse and his load were swept away, but the name of Slough Creek remains."[15]

It is mainly appropriate, for most of the stream meanders though lush meadows where — in periods of low flow — much water passes underground, leaving a series of isolated pools and segments. But in times of heavy runoff there is a roiling flood — particularly in the four-mile channel below the "Transfer."

Spinoffs of this name are few. Slough Creek Ranger Station appeared on the Park-wide topographic map in 1928, and it was still shown on the 1954 revision although the station had been gone a number of years by that time. The administrative uses appearing on current maps are Slough Creek Patrol Cabins (1959), a name designating several old buildings above the "Transfer" shown on earlier maps as "Hay Ranch and Patrol Cabin"; and Slough Creek Campground at the end of the auto road below the "Transfer."

Slough Creek Transfer — an informal name — denotes a particularly difficult section of the old wagon road that once connected Silver Tip Ranch, north of the Park boundary, with the Park road system. Prior to World War II, anything going into the ranch was actually transferred at that point to vehicles that had been laboriously brought over the bad section for use in traversing the meadows above. Use of the road above the campground was limited to horse-drawn vehicles after the War, and now with the ranch accessible by air, it has ceased entirely. This colorful name will undoubtedly fade away even faster than the unused road up Slough Creek.

Before going on to another place name, a few words are in order concerning a bit of local legend that attempts to account for the Slough Creek name in a ridiculous manner. In his little booklet on the Silver Tip Ranch, A. A. Tony Bliss says:

> Slough Creek has always seemed to me a shocking misnomer. There are sloughs here and there in the meadow lands, but they by no means justify the name. Actually, it was christened so unfortunately because a trooper of Troop 1 [*sic*], 2nd U.S. Cavalry, got lost in the valley in 1873. His name was John P. Slough.[16]

As already mentioned, Slough Creek received its name from a party of prospectors in 1867, and that usage was confirmed by the United States Geological Survey of the Territories, which showed it on the map appearing opposite page 162 in its *Fifth Annual Report* (1872). Public acceptance is evident from inclusion on the *Map of Montana, Eastern Portion*, published by Asher and Adams in 1873. The Blisses probably received their erroneous information from Superintendent Horace M. Albright, who had it from old soldier John P. Slough himself in 1922.

Trumpeter Slough of Troop I, 2nd Cavalry, was serving as escort for the Jones Expedition, which explored Yellowstone Park in August of 1873. His letter states:

> We attempted to follow the Hayden trail to ascertain where he had crossed the Yellowstone River below the Lake. We lost the trail and my Captain Henry E. Noise [Noyes] and my self started out to find it and in the timber and mountains [Mirror Plateau] we became lost and on the second day were found by a detachment under Sergeant Kinnoss, and the creek found upon now bears my name (Slough Creek).[17]

Note that Slough does not *say* the stream was named for him; only that it "now bears my name," which, strictly speaking, does not account for *how* the name came into being. In his second letter, covering his Yellowstone experience in greater detail, Slough Creek is not mentioned. It would seem that A. A. Bliss made an unwarranted assumption in attributing the Slough Creek name to John P. Slough.

## SODA BUTTE

This usage, which is of an undeniable prospector origin, is just as certainly based on a false premise.

Soda Butte, a striking hot spring cone about two and a half miles above the mouth of Soda Butte Creek, is the feature that suggested this place name. It was given by A. Bart Henderson, who had descended Cache Creek to the Lamar River with a prospecting party when he made this entry in his diary:

> Camped on East Fork, south side. Just opposite camp a small creek empties into river. One mile up this creek is a very singular butte, some 40 feet high, which has been formed by soda water. We gave the cone the name of Soda Butte, & the creek the name of Soda Butte Creek.[18]

Henderson was never an accurate observer, and in this case he ran true to form; Soda Butte was twice as far from his camp but only half as high as he made it, and it was not formed of soda ash (sodium carbonate) as he thought. It is really a deposit of travertine (calcium carbonate) — the same as Mammoth Hot Springs.

Soda Butte. A prominent cone of travertine (calcium carbonate) named by prospector A. Bart Henderson in 1870. He thought it was composed of soda ash (sodium carbonate). Photograph by William H. Jackson in 1872.

Regardless of the chemical inaccuracy of Soda Butte, it had staying power as a place name. The map that recorded the Hayden Survey's 1871 season in the Park designated Soda Butte Creek as "Sulphur Spring Creek," and Barlow's map used "Buffalo Creek," but neither survived. By the time the Geological Survey took over mapping in the Park, common usage had decided the issue, and all official maps from 1885 to the present have labeled the hot spring cone "Soda Butte" and the adjacent stream "Soda Butte Creek."

There have been a number of spin-offs from the basic place name, but none have survived. Superintendent Norris used the term "Soda Butte Medicinal Springs" for the seepage of warm water on the creek side of Soda Butte in 1879, and he shortened that to "Soda Butte Springs" in 1880.[19] Although Norris was lavish in his praise of the springs at Soda Butte, they were not prominent enough to attract further notice.

An officer who accompanied General Philip H. Sheridan on his monumental junket across Yellowstone Park in 1881 recorded another name that has not survived — "Soda Butte Lake." From his remark, "It is said that the miners of Cook City took ten tons [of fish] out of it last fall by drawing the lake and using nets, and dried them for winter's use,"[20] we recognize present Trout Lake.

Also in the limbo of lost place names is "Soda Butte Mail Station," established in 1884 on the road to Cooke City, and the "Soda Butte Station." The latter was occupied by a detachment of soldiers in 1886 and continued in use as a ranger station until the building was moved to the Buffalo Ranch prior to 1938 (although the station was still shown at Soda Butte on the 1954 revision of the Park-wide map).

The last of the Soda Butte names that failed to survive was "Soda Butte Canyon." The United States Geographic Board published a decision on May 7, 1930, assigning the name "Soda Butte Canyon" to the entire valley of Soda Butte Creek from the Soda Butte nearly to Silver Gate.[21] Another decision in 1933 limited Soda Butte Canyon to a small, roadside gorge a mile above Round Prairie. But that was later changed to Ice Box Canyon, and the Soda Butte Canyon name never got on the map.[22]

## TURKEY PEN

In this usage, so reminiscent of the Appalachian Mountains, is a touch of prospector humor.

It originated in the building of a cabin at the point of first settlement in what is now Yellowstone Park. George A. Huston, who was then interested in the placer at the mouth of Bear Creek, established himself in 1867 or 1868 at a spring close to the foot of Mount Everts and west of the intermittent stream now known locally as Turkey Pen Creek. Legend explains the name this way: Huston was raising the wall logs for the cabin when he was visited by several prospector friends. "What you doin', George?" asked one. "Puttin' up a cabin," he answered. The interrogator took a long look at the unchinked poles and replied, "Huh, looks like a turkey pen!"

George Huston, John Evans, and a man called Groves were living in the Turkey Pen Cabin when Jack Baronett brought the rescued Truman C. Everts there in 1870 to be nursed by the miners until transportation to Bozeman could be managed.[23] It is not known when the Turkey Pen Cabin was abandoned, but some occupational debris still marks the spot.

The place name — Turkey Pen Creek — was in use at least as long ago as 1886. It appears in that year in an account of the rescue of Truman C. Everts written by G. L. Henderson,[24] yet it has never appeared on official maps of the Park (locally, it is still known and used).

Turkey Pen Trail — which could even be called a road, as freight was hauled over it as early as 1878 — is the designation formerly applied to the old miner's route that connected the mouth of Gardner River with the Cooke City mines. It was named that on the 1912 Park-wide map of the Geological Survey, but present maps call it Rescue Creek Trail.

The only place name from this root remaining on official maps of the Park is Turkey Pen Peak. In 1955 the Geological Survey gave that name to a prominent point east of the head of "Turkey Pen Creek." Thus, a place name that is both colorful and historic is not entirely lost.

## DELACY

While not a usage established by a prospector, this one remembers a man who *was* a prospector — and much more — and certainly deserved the honor.

Both of the General Land Office maps of 1869 show a "DeLacy's Lake" at the head of the Snake River. On the earlier of the two, it is too far west of Yellowstone Lake. The later version, which benefited from the information brought back by the Folsom party that fall, moved DeLacy's Lake more nearly into a correct relationship, though placing it a little too far south in order to accommodate an unnamed triangular lake thought to be the source of Firehole River.

DeLacy Lakes. Walter Washington deLacy piloted a party of prospectors across the SW corner or Yellowstone Park in 1863, correctly mapping the drainage pattern. These lakes at the head of DeLacy Creek honor his early mapmaking.

This fictitious "Madison Lake," as it came to be called, was really DeLacy's Lake seen from the north but appearing much different than it had to Walter W. deLacy in 1863 (he had passed up the east shore and thus was able to see the westward expanse of the lake).

As mentioned in Chapter 1 (see Shoshone), this unfortunate accumulation of mapping errors led Professor Frank H. Bradley to deny deLacy's right to have his name perpetuated on the lake he had been first to correctly relate to the Snake River drainage. As a consequence, DeLacy's Lake was given another name by which it has been known ever since — Shoshone.

Superintendent P. W. Norris was disturbed by the failure of Ferdinand V. Hayden to correct the wrong done by his assistant. Seeing no likelihood of a restoration of deLacy's name to the lake, he "as a small token of deserved justice, named the stream and park crossed by our trail above the Shoshone Lake after their discoverer."[25]

The Geological Survey put DeLacy Creek on the map in 1885, and in 1956 — as an added measure of justice — a group of previously unnamed lakes on the Continental Divide at the head of DeLacy Creek were designated "DeLacy Lakes."

Administrative usage has added the name "DeLacy Creek Trail" to identify the trail that connects the outlet of Shoshone Lake with the Howard Eaton Trail where the latter descends to DeLacy Creek east of Norris Pass.

## MISCELLANY

Of ten isolated place names that can be traced to the prospectors, only two have survived to this time — and one has a poor pedigree.

The use of "Great Falls" to distinguish the double falls of Yellowstone River from the "Little Falls" on Tower Creek was recorded by A. Bart Henderson in 1866.[26] Steamboat Springs, at Steamboat Point on the east shore of Yellowstone Lake, goes back to an 1867 newspaper item that mentions "a steamboat on the side of the big lake above here [Bear Creek] that shoots puffs of smoke out of a volcanic peak over the bosom of the waters."[27] The Hayden Survey called that place Steam Point and Steamboat Point in 1871, and the Geological Survey has always used Steamboat Springs for the thermal feature and Steamboat Point for the locale.

"Alum Creek" appeared on the two maps forwarded by the surveyor general of Montana Territory to the commissioner of the General Land Office in 1869, and was also shown on the 1870 edition of deLacy's map. On all these maps it is a stream entering Yellowstone River from the east about where Broad Creek ought to come in. This place name could be accounted for as a product of the Folsom party's exploration of the Yellowstone country in 1869 except that it occurs on the earlier map without the other geographical information brought back by David E. Folsom. Thus it would appear to predate that exploration, and with no ties to the trapper period, it is here arbitrarily relegated to the prospectors. This place name continues to be a problem in Chapter 4.

On October 1, 1869, David E. Folsom used the name "Death Valley" in his diary to describe the Lower Geyser Basin, which he recognized from the descriptions of "persons who had previously visited it."[28] Those persons could have been George Huston, who talked that way after his retreat from the geyser basins in the late fall of 1864, or David B. Weaver, whose letter to the *Montana Post* in 1867 mentioned the experience of a party of prospectors: "The atmosphere was intensely suffocating, and they report that life could not long be sustained there. Not a living thing, bird or beast, was seen in the vicinity."[29] The prospectors thus left a legacy of fear concerning the geyser basins to haunt the early years of the Park. Even Superintendent Norris aborted a stay in the Upper Geyser Basin from fear of the deleterious effects of a vapor-laden winter climate.

Another place name that can be attributed to A. Bart Henderson is "Buffalo Flat," a name given on June 21, 1870, because "we found thousands of buffalo quietly grazing."[30] If we amend Henderson's claim to read "hundreds," it would be about right; nevertheless, the Geological Survey perpetuated his usage as Buffalo Plateau.

Henderson and three of his rough companions, Adam "Horn" Miller, Ed. Hibbard, and James Gourley, prospected the country at the head of the Upper Lamar River in 1870. It was this party who discovered the "region of countless remnants of erosion, so wild, weird, and spectral that they named it the 'Hoodoo' or 'Goblin Land'"[31] before they were driven out of the area by hostile Indians. From that characterization came two place names in use on current maps: Hoodoo Basin and Hoodoo Peak.

"Warm Spring Creek" is a name on which the prospectors didn't always agree. Most often it was used to designate present Tower Creek — from the tepid springs at its mouth — but that stream was also called "Little Fall Creek" by some, and just "Fall Creek" by others. If that wasn't confusion enough, they also referred to Gardner River as "Warm Spring Creek," "Warm Stream Creek," and "Hot Spring Creek" — but then, in such matters each prospector was his own authority.

There is one name that does not really belong to the period of the prospector, but it immortalizes a man who might be considered typical of the breed. The same year that John Yancey sold his mining property on Crevice Mountain, the Bozeman newspaper mentioned "John Knowles, of Crevice Gulch."[32] He and others seem to have continued mining operations at the mouth of Crevice Creek until May 24, 1898, when the soldiers burned all the cabins except one that had been occupied by Knowles. That building, thereafter called Knowles Station, was used by army patrols charged with keeping the miners and illegal fishermen out.

During the years Knowles worked his placer at the mouth of Crevice Creek he had access to it by a trail down Blacktail Deer Creek from the Cooke City road. It ran essentially along the present trail, which descends to the suspension bridge, except that he had to cross the Yellowstone River at the ford a half mile above. That trail was shown on the 1912 Park-wide map as "Knowles Trail," which hints that he was its builder.

The creek down which Knowles built his trail — Blacktail Deer Creek — probably bears a name of prospector origin, yet it escaped notice until the Barlow party came along in 1871 (there is nothing to indicate they named it; it just appears on their map). Regardless of the name's lack of pedigree, the Geological Survey has never questioned its use on Park maps. A related place name, Blacktail Deer Plateau, was given to the expanse of beautiful grassland with intervening groves of aspen and fir that extends eastward as far as The Cut.

John S. Knowles, the average prospector-miner, has been immortalized through the application of his name to a waterfall on the opposite side of the Yellowstone River about a quarter mile below the site of his cabin (1955 topographic map, Gardiner quadrangle).

Earlier I alluded to the prospector's penchant for sometimes giving his place names a ribald slant. An entry in the diary of William Blackmore, the English anthropologist who was a guest of the Hayden Survey in the summer of 1872, provides a typical example. Blackmore accompanied Hayden and Holmes on a side trip to the "miner's camp" at the head of Soda Butte Creek (the vicinity of present Cooke City). While there, he learned the local name for one of the spectacular summits that crown the south wall of the valley (probably Pilot Peak). With Victorian delicacy, Blackmore entered the name in Pitman shorthand, which translates into very plain English as "dog turd mountain."[33]

A similar reluctance concerning the crude usages of the mountaineers is latent in Lord Dunraven's comment concerning the pillar of an extinct hot spring that stands, phallus-like, at the foot of Mammoth Hot Springs (named "Liberty Cap" by the Hayden Survey in 1871 from a fancied resemblance to the peasant cap popularized in the French Revolution). Among the things Dunraven took note of on his visit to the Springs in July, 1874, was that pillar: "Professor Hayden calls this the Liberty Cap; locally it goes by another name."[34] What that was he discreetly left to our imagination.

A place name that developed out of a prospector's "spoof" to be temporarily associated with the Yellowstone region will do to close this chapter. Though fraudulent, it had the good effect of initiating the definitive exploration of the area that is now Yellowstone National Park.

Under the lengthy byline, "Good Story — We clip from an eastern paper the following, which will do to counter-balance some of the fearful lies that have been told for the purpose of showing that the country was played out," the editor of a Helena newspaper laid this one on his readers:

> Mr. Edward Parsons, just returned from Montana, tells the editor of the Leavenworth *Commercial* a marvelous story. Last July, himself and four companions, while exploring the headwaters of the Yellowstone, came upon an Indian mound, surmounted by a huge stone. Dislodging this stone and several others, they found themselves in an Indian catacomb, containing the skeletons of thirty warriors. Lying beside the bones were numerous ornaments, among them many twisted circlets of gold. Some of these were of unusual size, weighing one and a half to two pounds. What chiefly attracted attention was a massive basin or kettle that occupied the centre of the apartment. This massive article proved to be pure gold, and was so heavy that the party had great difficulty in removing it from its resting place and bringing it into the upper air. The adventurers were enabled, by means of their axes, to sever the mass into portable pieces, laden with which, the party turned their steps homeward, having themselves to walk the greater part of the way, to give relief to their burdened animals. The whole amount of gold was brought to Helena, and Mr. Edward Parsons calculated that his share of the treasure amounted to about $21,000, the whole bulk being at least $100,000 in value.[35]

Of course, the story was ridiculous; it was the sort of thing frontier editors ran as entertainment. Yet it generated a myth of "the mysterious mounds" somewhere in the Yellowstone region. A year and a half later the same newspaper ran this notice:

> *Expedition to the Yellowstone.* A letter from Fort Ellis, dated the 19th, says that an expedition is organizing, composed of soldiers and citizens, and will start for the upper waters of the Yellowstone the latter part of August, and will hunt and explore for a month or so. Among the places of note which they will visit, are the Falls, Coulter's Hell and Lake, and the Mysterious Mounds. The expedition is regarded as a very important one, and the result of their explorations will be looked forward to with unusual interest.[36]

At Diamond City, a mining town in Confederate Gulch across the valley from Helena, three men read that notice and decided they would like to go along. By the time they inquired, however, the expedition had been cancelled from fear of Indians. The three friends, David E. Folsom, Charles W. Cook, and William Peterson thus proceeded on their own, and the information brought back by this Folsom Expedition of 1869 opened the period of definitive exploration of the Yellowstone region.

I like to think that interest in the "Mysterious Mounds — a thought that there *might* be something to that tale — was the crucial factor in the decision to go, and thus set in motion the events that led quickly to establishment of our first national park. If so, those nonexistent "Mysterious Mounds" were a bridge between the prospectors and the explorers who followed them.

## NOTES

1. Lewis Carroll (Charles Lutwidge Dodgson) used the word in the title of his children's book, *Alice's Adventures In Wonderland* (Oxford University Press, 1866), thereby bringing it to literary prominence.
2. A. Bart Henderson, *Journal of the Yellowstone Expedition of 1866 Under Captain Jeff Standifer . . . Also the Diaries Kept by Henderson During His Prospecting Journeys in the Snake, Wind River, and Yellowstone Country During the Years 1866–72.* Transcription by Granville Stuart, Coe Collection Item No. 452, Beinecke Library, Yale University, New Haven. Both excerpts are from p. 107.
3. C. C. Clawson, "Notes on the Way to Wonderland," *Deer Lodge* (MT) *New North-West,* September 9, 1871, through February 24, 1872 (Serially — 12 issues under that title). Also, "The New Wonderland," *New York Daily Times,* October 23, 1871.
4. Philetus W. Norris, *Fifth Annual Report of the Superintendent of the Yellowstone National Park,* Dec. 1, 1881 (Washington: Government Printing Office, 1881), p. 44. The other account is by Eugene S. Topping, *The Chronicles of the Yellowstone* (St. Paul, MN: Pioneer Press, 1883), pp. 24–25. In it the leader of the original party is identified as George A. Huston, which is correct, and the Indian horse thieves are called Arapahoes, which is unlikely.
5. John W. Davis, "Yellowstone Park," *Louisville* (KY) *Courier-Journal,* April 18, 1884, p. 12.
6. Ferdinand V. Hayden, *Yellowstone National Park,* 1871. National Archives, RG–57, Hayden Survey, YNP.
7. Hiram M. Chittenden, *Yellowstone National Park* (Cincinnati: Robert Clarke, 1915), p. 162.
8. [Editorial] Virginia City (MT) *Montana Post,* July 14, 1866.
9. Manuscript maps of the Territory of Montana that accompanied the report of the surveyor general for 1869. National Archives, RG–49, GLO(OMF), Montana 2 & 3.
10. Ibid., Montana 2.
11. Topping, *Chronicles,* pp. 62–63; see Note 4 for citation.
12. William Blackmore, *Diary — 1872 — Fourth Visit to the United States,* being Parts 5, 6 & 7 of the MS diary of William Blackmore. Transcript in the Yellowstone Park Reference Library, Part 6, p. 71.

13. "The Mines of Southeastern Montana," Bozeman (MT) *Avant Courier*, February 12, 1880.
14. Topping, op. cit., p. 63.
15. Ibid.
16. A. A. Tony Bliss, *Silver Tip Ranch, 1922–47* (privately printed, 1947), p. 4.
17. John P. Slough, Letter to Superintendent Horace M. Albright, August 6, 1922. *Cf.*, his second letter (January 13, 1924). Yellowstone Park Reference Library.
18. Henderson, *Journal*, p. 93 (July 23, 1870); see Note 2 for citation.
19. Philetus W. Norris, *Report Upon the Yellowstone National Park, to the Secretary of the Interior, for the Year 1879* (Washington: Government Printing Office, 1880), pp. 23 & 30. Also, *Annual Report of the Superintendent of the Yellowstone National Park, to the Secretary of the Interior, for the Year 1880* (Washington: Government Printing Office, 1881), map.
20. Philip H. Sheridan, "Report of Lieut. Col. James F. Gregory," *Report of Lieut. General P. H. Sheridan, Dated September 20, 1881, of His Expedition Through the Big Horn Mountains, Yellowstone National Park, Etc.* (Washington: Government Printing Office, 1882), p. 18.
21. *Decisions of the United States Geographic Board, Yellowstone National Park* (Washington: Government Printing Office, 1930), p. 21.
22. *Decisions of the United States Board on Geographic Names, July 1, 1936 to June 2, 1937* (Washington.: Government Printing Office, 1937), p. 15.
23. Truman C. Everts, "Thirty-seven Days of Peril," *Scribner's Monthly*, 3/1 (Nov. 1971), p. 16. The cabin was identified by name in 1883 (Topping, *Chronicles* [1883], p. 84).
24. George L. Henderson, manuscript journal kept at Mammoth Hot Springs, 1882–1886, Yellowstone Archives, Item No. 141, p. 100. See July 1, 1886.
25. Norris, *Fifth Annual Report* (1881), p. 44.
26. Henderson, *Journal*, p. 93; see Note 2 for citation.
27. "Interesting from the Yellowstone," Helena (MT) *Daily Herald*, Apr. 11, 1867.
28. Charles W. Cook, David E. Folsom, and William Peterson, *The Valley of the Upper Yellowstone*, ed. by Aubrey L. Haines (Norman: University of Oklahoma Press, 1965), p. 42.
29. "D. W." [David B. Weaver], "The Upper Yellowstone," Virginia City (MT) *Montana Post*, August 31, 1867.
30. Henderson, *Journal*, p. 81; see Note 2 for citation.
31. Norris, *Report* (1880), p. 7; see Note 19 for citation.
32. "Newsy Letter from Upper Yellowstone," Bozeman (MT) *Avant Courier*, April 6, 1882.
33. Blackmore, *Diary — 1872*, No. 6, p. 88; see Note 12 for citation.
34. Windham T.W.Q. Dunraven, *The Great Divide . . . Travels in the Upper Yellowstone in the Summer of 1874* (London: Chatto and Windus, 1876), p. 194.
35. Helena (MT) *Weekly Herald*, Dec. 26, 1867, p. 7.
36. Helena (MT) *Weekly Herald*, Jul. 29, 1869, p. 7.

# 4

## GIVEN BY THE EXPLORERS



The period of exploration of the Yellowstone wilderness differs from those already considered in that it is not a cohesive time span, but several widely separated episodes related only by their motivation — geographical curiosity. These can be grouped as related to the Lewis and Clark Expedition's journey to the Pacific Ocean and return in 1804–1806, to the interest of the Army Corps of Topographical Engineers in the headwaters of the Yellowstone River in 1859–1860, to the definitive exploration of the upper Yellowstone country by the Folsom, Washburn, Hayden, and Barlow parties in 1869–1871, and to the Jones party in 1873.

The earliest explorers, the men of the Lewis and Clark Expedition, were forty-three miles north of what is now Yellowstone National Park at their closest approach (while returning from the Pacific). Their contribution to knowledge of that area — prior to John Colter's return to civilization in 1810 — amounted only to identification and naming of several rivers whose headwaters are on the Yellowstone plateau. As explained in Chapter 2, John Colter's activity as a trapper after leaving the Lewis and Clark Expedition resulted in a winter journey across part of the present Park, and some of his observations were eventually published with the Lewis and Clark report.

Another exploration contemporaneous with that of Lewis and Clark — the 1806 examination of the Arkansas headwaters by Zebulon Pike — planted the idea that the western plains acquired with the Louisiana Purchase were a "great American desert." Stephen Long's reiteration of that thesis in 1820 resulted in official recognition of a "permanent Indian frontier" — an intent to leave the plains in the hands of the Indians. Consequently, there was a lapse in official exploration until the 1840s, when the overland migrations to the far west renewed interest in western geography.

John C. Fremont of "Pathfinder" fame launched the Army Corps of Topographical Engineers into a program of exploration that provided maps for many areas of the West. In 1856 Lieutenant G. K. Warren began the mapping of the Yellowstone River, and he had pushed the work up to the mouth of the Powder River when it was taken over by Captain W. F. Raynolds in 1859. Raynolds, hampered by strict scheduling combined with the late-lying snows of the winter of 1859–1860, was prevented from entering what is now Yellowstone Park, and the opening of the Civil War delayed publication of his map and report for a decade. By then they were of minimal value, although they did perpetuate some of Jim Bridger's information about the Yellowstone country.[5]

Raynolds's only contribution beyond what he got from his guide was the sobriquet, "Valley of the Upper Yellowstone" — a term with a connotation of awe and mystery similar to "Wonderland" at a later date.

Definitive exploration began with the Folsom party in 1869, continued with the Washburn party of 1870, and was brought to completion by the Hayden and Barlow parties in 1871. The first of these definitive explorations, a fall visit by three miners from Diamond City, Montana Territory — David E. Folsom, Charles W. Cook, and William Peterson — was prompted solely by curiosity. These men named few features, with only two of those surviving to the present. Yet their visit was not without merit in a cartographic sense, for the information with which Folsom returned (he was a surveyor by training) allowed Walter W. deLacy to revise his *Map of the Territory of Montana* (1870) so that it portrayed the Yellowstone region with reasonable accuracy — for the first time. Beyond that, their accurate account of the region through which they had passed appeared in magazine form within a year (if the published account did not reach Helena prior to their departure, the explorers of 1870 already had the gist of it directly from Folsom, who was by then employed in Surveyor-General Washburn's office).

Thus, the larger expedition into the Yellowstone in the summer of 1870 had access to map information in addition to the accounts of Folsom and Cook. The personnel of this "Washburn Party" consisted of twelve civilians under the leadership of Henry D. Washburn and a military escort of four cavalrymen commanded by Lt. Gustavus C.

Doane — a total of seventeen. It was a semiofficial expedition, well equipped and well led, and so was able to obtain much more information than the Folsom party could with their limited means. However, their accomplishments included more than identification and naming of features, for the publicity that followed — newspaper accounts, magazine articles, and lectures — served to strip away the veil of mystery that had obscured the true nature of the area. Before 1870 only a handful of people knew what lay "behind the ranges"; thereafter, anyone who could read could know.

The final phase of definitive exploration of the Yellowstone region, in 1871, was the work of two government organizations: the United States Geological Survey of the Territories, better known as the "Hayden Survey," and a small party under two officers of the Army Corps of Engineers, captains John W. Barlow and David P. Heap. These parties traveled in tandem, sharing a military escort, and they were able to augment the largely descriptive accounts of the two previous explorations with collections and measurements of a scientific nature, better cartography, and — most convincing of all — the first photographs.

Thus the definitive exploration that began with the Folsom party in 1869 and continued with the Washburn party in 1870 was completed by the Hayden and Barlow parties in 1871. The effect was to provide an incontrovertible knowledge of the nature of the Yellowstone wilderness, and to establish the factual basis for a movement to create a *national* park dedicated to preservation of the region's thermal and scenic wonders "for the benefit and enjoyment of the people."

But the period of the explorers did not end neatly with the definitive exploration of the Yellowstone region. There was a lingering interest by the Army Corps of Engineers (the post–Civil War counterpart of the Corps of Topographical Engineers), which sent Capt. William A. Jones on one last, grand exploration in 1873 — two years too late. The Hayden Survey was already routinely filling in the chinks in the map.

## GALLATIN

This place name was used by Lewis and Clark on July 28, 1805, to honor Albert Gallatin, secretary of the treasury. As the easternmost of the three forks that combine to form the Missouri River, it drains the west flank of the mountain range occupying the northwest corner of Yellowstone Park.

Concerning the naming of the streams encountered at the Three Forks of the Missouri, Captain Lewis wrote:

> Both Capt C. and myself corrisponded in opinion with rispect to the impropriety of calling either of these streams the Missouri and accordingly agreed to name them after the President of the United States and the Secretaries of the Treasury and State having previously named one river in honor of the Secretaries of War and Navy. In pursuance of this resolution we called the S.W. fork, that which we meant to ascend, Jefferson's River in honor of that illustrious personage Thomas Jefferson (the author of our enterprise), the Middle fork we called Madison's River in honor of James Madison, and the S.E. Fork we called Gallitin's River in honor of Albert Gallitin.[1]

Albert Gallatin's surname was similarly misspelled on the manuscript map prepared by Captain Clark (his spelling was unconventional at the best), and it remained for Samuel Lewis and Nicholas Biddle to put that place name straight for posterity.[2]

Spin-offs from this place name are a short-lived "Mt. Gallatin," which appeared on the map accompanying the report of Captain Raynolds,[3] but was renamed Mount Holmes by the Hayden Survey in 1878; Gallatin Range, thought by historian Chittenden to have originated with the prospectors of 1870; Gallatin Lake, named by the Geological Survey in 1885; two administrative usages — Gallatin Station (1912; became Gallatin Ranger Station in 1928) and Gallatin Petrified Forest on the 1955 revision of the Crown Butte (Montana) quadrangle.

## MADISON

Another of the three place names given by Lewis and Clark on July 28, 1805, to identify streams forming the Missouri River at Three Forks, Montana, Madison identifies a river that drains most of the west-central area of Yellowstone Park, including the most important geyser basins.

Appearing first as Madison's River, for Secretary of State James Madison, the name was used in the possessive form until mid-century. Captain Raynolds seems to have been the first to use it as spelled today.[4] His map also showed a "Mt. Madison" south of the headwaters of Madison River, a feature for which we cannot logically account.

Madison River. The Lewis and Clark Expedition named this river in 1805 after Secretary of State James Madison. It drains the central part of the Park's west side, including the geyser basins of the Firehole and Gibbon Rivers.

The first important variation to develop from the original usage was Madison Lake; or rather, the "Madison Lake" (present Shoshone) that appears on the 1870 edition of the deLacy map.[5] This probably resulted from David E. Folsom's misconception that the lake he saw in 1869 was the source of Firehole River and thus the headwater of the Madison system.

That cartographic error was accepted by the Hayden Survey and repeated on the preliminary map produced in 1871,[6] with a surprising result. Hayden's map became the basis of the boundary proposal submitted to Congress for use in drawing up the legislation that established Yellowstone National Park on March 1, 1872. That proposal established the Park's west boundary by prolonging the south boundary "to the meridian passing fifteen miles west of the most western point of Madison Lake; thence north."[7] The Park's western line was not surveyed until many years after the Madison name had been shifted (in 1872) to a pond that really *was* at the source of that drainage.[8] The surveyors, though aware of the change, placed their reference point on the west shore of Shoshone Lake (1894), where the stone monument can still be seen. This was done on the basis of a decision in the office of the secretary of Interior to survey the Park's western boundary in accordance with the *intention* of the act of dedication (see *Annual Report, 1894;* p. 5).

There is an interesting reference to the pond that now bears the name of Madison Lake. In the 1903 edition of *The Yellowstone National Park,* historian Chittenden notes: "This body of water, with the exception of Red Rock Lake, the source of the Jefferson, is further from the sea by direct water course than any other lake on the globe" (p. 301). While I am not aware of anything that confirms this claim, Chittenden's seven years of work with the Corps of Engineers on improvement of the Missouri River should have put him in the position of an authority on the world's waterways.

Other variations that have developed from the original usage over the years include Madison Creek, another Hayden Survey name of 1871, appropriately changed to DeLacy Creek by Superintendent P. W. Norris in 1881,[9] and Madison Plateau.

This latter place name as originally applied by the Hayden Survey (1878) included most of the west side of the Park, but the Geological Survey (1885) reduced its extent to the area south of Madison Canyon, as shown on present maps. The map accompanying Superintendent P. W. Norris's *Annual Report* for 1880 has a "Madison Plateau Road" between Riverside Mail Station (on Madison River) and the mouth of Nez Perce Creek. This attempt to avoid the many stream crossings on the road through Madison Canyon was so steep where it climbed onto the plateau that it was often referred to as the "Norris Slide." Though a failure as a road, the route is still used as a trail.

This is the proper place to mention that the area first known as the "Burnt Hole," the circular valley that holds Hebgen Lake and the town of West Yellowstone, received its present designation — Madison Valley — in 1915. As a place name, then, the terminology belongs *only* to a limited area, not the entire course of the river. Similarly, the practice of earlier days of extending the Madison name up the various forks including "South Fork" (Firehole River since 1871), "East Fork" (Gibbon River since 1873), and "East Fork" (Nez Perce Creek since 1885) was confusing. The appearance of Madison Junction as a place name on the 1928 revision of the Park topographic map served to accent what had been a fact for many years — that the Madison River began there!

Two administrative usages complete the spin-offs from the Madison name: Madison Range Overlook and Madison River Lookout. Both appeared on the topographic quadrangles (Madison Junction and West Yellowstone) in 1958; however, the latter was not a new place name. It had been used from early in the 1930s to designate an emergency fire lookout station, though not shown on the map previously.

## TOWER

This place name, which was the Washburn party's first important one within what is now Yellowstone Park, was given to the beautiful waterfall known to the trappers as "Little Falls" and to later prospectors as "Lower Falls" — both of those in contradistinction to the "Great Falls" of the Yellowstone River.

The rudimentary diary of Sam Hauser indicates that the name Tower Falls was his idea. On August 27, 1870, he wrote:

> Stil travling in foot Hills-descended mountains to the river.
>
> Campt near the most beautiful falls-I ever saw-I named them 'Tower Falls'-from the towers & pinacles that [overhang] them height 115 feet Aug 28-Didnt move camp Cenery too beautiful I measured these falls-found them 115-the pinacal cover top Then 200 ft Eight different spires [?] crown the falls-From below the view is splendid[10]

Nathaniel P. Langford, in his published *Diary*, has much to say about the naming of this feature (though the reader will discover he is not to be taken seriously). Langford begins: "In camp today several names were proposed for the creek and fall, and after much discussion the name 'Minaret' was selected. Later, this evening, this decision has been reconsidered, and we have decided to substitute the name 'Tower' for 'Minaret,' and call it 'Tower Fall.'"

This is fine until one takes up the footnote, which proposes to be "an account of the circumstances under which the name 'Tower' was bestowed." Continuing from the footnote:

> At the outset of our journey we had agreed that we would not give to any object of interest which we might discover the name of any of our party nor of our friends. This rule was to be religiously observed. While in camp on Sunday, August 28th, on the bank of this creek, it was suggested that we select a name for the creek and fall. Walter Trumbull suggested 'Minaret Creek' and 'Minaret Fall.' Mr. Hauser suggested 'Tower Creek' and 'Tower

Fall.' After some discussion a vote was taken, and by a small majority the name 'Minaret' was decided upon. During the following evening Mr. Hauser stated with great seriousness that we had violated the agreement made relative to naming objects for our friends. He said that the well known Southern family — the Rhetts — lived in St. Louis, and that they had a most charming and accomplished daughter named 'Minnie.' He said that this daughter was a sweetheart of Trumbull, who had proposed the name — her name — 'Minnie Rhett' — and that we had unwittingly given to the fall and creek the name of this sweetheart of Mr. Trumbull. Mr. Trumbull indignantly denied the truth of Hauser's statement, and Hauser as determinedly insisted that it was the truth, and the vote was therefore reconsidered, and by a substantial majority it was decided to substitute the name 'Tower' for 'Minaret.' Later, and when it was too late to recall or reverse the action of our party, it was surmised that Hauser himself had a sweetheart in St. Louis, a Miss Tower. Some of our party, Walter Trumbull especially, always insisted that such was the case. The weight of testimony was so evenly balanced that I shall hesitate long before I believe either side of this part of the story.[11]

Tower Fall. Sam Hauser of the 1870 Washburn party named and measured this waterfall. Those explorers worried that the large rock at the top was about to fall, but it remained there until 1990 — another 120 years!

Tower Falls was the form in which this place name was used until the Geological Survey changed it to the singular — Tower Fall — on its 1928 revision of the Park-wide topographic map. That usage has persisted since, and properly so, as the waterfall consists of one clean drop of 132 feet.

The importance of Tower Fall as a Park feature following completion of the Park's figure-eight road system (1903) led to an increased use of that place name to identify features in the vicinity. The new road to the Grand Canyon area was soon known as the "Tower Falls Road" as far as the Dunraven Cutoff, and the military outpost established to control the traffic became "Tower Falls Station," though not located at Tower Fall but north of it. This station was first located between Overhanging Cliff and Rainey Lake, and later (1907) at the new junction with the road to Cook City.

A further expansion of the Tower usage took place with the appearance of "Tower Junction" and "Tower Ranger Station" on the 1959 revision of the topographic map (Tower Junction quadrangle). The name was thus confirmed in a regional connotation that had been implied by local usage after 1907.

## WASHBURN

If the Washburn party did agree not to name features for themselves, as Langford indicated, it was a resolution soon broken by naming a peak for their leader, Henry D. Washburn. Regardless of this infraction of their good intensions, it was a good decision. Mount Washburn is a fitting memorial for a man whose life was not only useful beyond the ordinary, but was shortened by the rigors of his Yellowstone exploration.

The circumstance that occasioned the naming is noted in the diary of Cornelius Hedges: "Washburn & Williamson [a soldier of the escort] have been out to find a trail up west side to save crossing river returned at night, reporting having seen Lake. The route practicable. scenery fine. resolved to go that way."[12]

The actual naming was done the following day, when five members of the expedition diverged from the Indian trail followed by the main party in order to visit the summit. Gillette's diary states:

> When we came to the dividing ridge Trumbull Hauser Langford Stickney & myself went up to the summit of the highest mountain which proved to be 10579 feet above the sea From the top of this mountain which we named Mt. Washburn a beautiful view was spread upon all sides of us."[13]

Hauser's diary entry for August 29 is similar: "Named highest Peak Washburn Peak. Langford Ben and I climbed to its top Barometer 10700 ft. high."

Henry D. Washburn died on January 26, 1871, taken by pneumonia that developed from a cold contracted on the Yellowstone expedition. Cornelius Hedges delivered a eulogy three days later in which he attempted to engraft the Washburn name onto Shoshone Lake in these words: "Between the Yellowstone Lake and the Great Geyser basin of the Madison, is also a beautiful sheet of pure, clear, trout haunted water [actually, barren of fish at that time] named Washburn Lake by Messrs. Folsom and Cook [not true, they called it Madison Lake] . . . the deep forests of evergreen pine, that embosom that lake, shall typify the evergreen spot in our memory."[14]

Other place names that have developed as spin-offs from the Washburn name are Washburn Hot Springs, a thermal area south of Mount Washburn (1904), Washburn Range, the rim of an ancient volcanic caldera of which Mount Washburn is the highest point (1930), and Mount Washburn Spur Trail (1959). The latter is an administrative usage for the link between the summit and the Howard Eaton Trail.

Mount Washburn (10,243 ft.). The highest point in the range seen across Hayden Valley was named by the Washburn party (1870) for their leader — Henry Dana Washburn, surveyor-general of Montana Territory.

## CASCADE

The naming of Cascade Creek, a beautiful stream entering Yellowstone River from the west between the Upper and Lower Falls, was the principal contribution of the Washburn party to the place names in that vicinity. Why it was named is clear enough. The last mile of the creek below the meadow in which the party encamped, descending as it does in a succession of rapids through a woodsy draw ("The Devil's Den" of those explorers), led naturally to that name.

Cornelius Hedges was the first to follow the stream down from the meadow to the beautiful falls immediately above the junction of the creek with Yellowstone River. Although Hedges doesn't use the name in his diary, Langford does: "Mr. Hedges gave the name 'Cascade Creek' to this stream."[15] Langford also credits Cornelius Hedges with the naming of Crystal Falls, another waterfall that is a most unusual, triple-stepped cascade.

Before taking up the spin-offs of this place name, it is necessary to point out that two other Yellowstone streams have been called Cascade Creek. The stream now known as Boundary Creek (1904) was originally "Cascade Creek," so named by Superintendent P. W. Norris in 1878.[16] A trace of that name remains on the 1957 Buffalo Lake quadrangle where it identifies a headwater northeast of Buffalo Lake. Another stream currently known as Cascade Creek is the one entering Yellowstone Park in the southwest corner, midway between Bechler and Snake Rivers.

Returning to spin-offs of the original Cascade Creek name, Photographer F. J. Haynes used "Cascade Falls" instead of Crystal Falls in his 1896 guide book — a usage that did not survive; but his "Cascade Creek Bridge" for the steel span just above the falls lasted until that bridge was replaced by an earth-fill crossing following World War II. Cascade Lake, the principal source of the creek, received its name from the Geological Survey in 1900 (appearing for the first time on the map in 1912), and the large meadow through which the creek passes for much of its course became Cascade Meadows in 1958 (Norris Junction quadrangle).

Other uses of the word "cascade" have been made in Park place names, and they will be considered where appropriate.

## MUD GEYSER/MUD VOLCANO

These two place names sound as if they might refer to the same feature, but they do not. Yet they have more in common than the "mud" in their labels: They were both given in a casual, uncertain manner, and both have survived to dominate that part of the thermal basin in which they are located, so that it has always been the "Mud Geyser area" or "Mud Volcano area" — a vague distinction indeed!

The Washburn party arrived in that vicinity the evening of September first, with most of the members in bad shape from an intestinal complaint variously attributed to the water and their canned peaches. In addition, Lieutenant Doane was suffering from a painful inflammation of one thumb. Those were likely the reasons the party laid over the following day, and it could have influenced the sloppy way they handled the naming of several impressive thermal features.

A beautiful, green-gabled spring expelling steam and clear water at the head of a shallow draw — present Dragons Mouth Spring — was variously noted as a "grotto" (by Doane), a "cave" (by Washburn), a "singular cave" (by Hedges), a "cavern" (by Langford), and as "Cave Spring" (by Hauser, Gillette, and Trumbull). This feature has also been called "Devil's Den" and "Devil's Workshop," but — fortunately — those visitor appellations did not become popular.

A hundred yards to the south a large vent in the hillside thumped noisily as it churned a steamy pool of mud, and evidence of more violent activity that had flung mud onto trees and the grass at a distance led Trumbull to call it a "volcano," with Hauser using "Mud spring," and the others (Doane, Langford, and Washburn) writing "a mud volcano." Only Hedges wrote "Mud Volcano" like he meant it as a name, and that was added separately at the end of his description.

A third feature at the foot of the hill, a funnel-like pool about thirty feet by forty, was observed to erupt in a powerful manner — the first geyser action the party had seen. It received mention merely as a "geyser" (by Hauser, Doane, and Trumbull), as "our first geyser" (by Washburn), as "muddy geyser" (by Langford, 1871), and as "water volcano," "great geyser," and "great Water Spout" (by Hedges). None of the writers of the expedition got it exactly right until Langford published his back-written *Diary* in 1905.[17]

## OLD FAITHFUL

Excepting the name Yellowstone itself, no other place name in the Park is as well known as the name of Old Faithful Geyser. As is usually the case with a popular name, it has been attached to many things and used for many purposes never imagined in the beginning — and it has even come to symbolize our great national park for many people.

The Washburn party's progress across the rough country between Yellowstone Lake and the Upper Geyser Basin was far from pleasant. Added to the misery of travel through an unknown forest, with snow on the ground and cold rain in the air, there was the worry over the fate of their comrade — Truman C. Everts — lost in the wilderness south of the lake, and personal worries concerning a dwindling food supply, worn out equipment and tired horses. Even more disquieting was the nagging question: Were they, too, lost? Cornelius Hedges summed up the general mood as the second day opened: "prospect was gloomy enough, felt double disgust. feet wet clothes wet . . . sat in the rain under tree moodily meditating."[18]

Old Faithful Geyser. The regularity of this geyser over the years (averages: interval, 65 minutes; eruption, 3 minutes; height, 130 feet) has made it the best-known feature in Yellowstone Park.

But that day turned out better than they expected. Hedges continued: "in about 2 hours by great good luck struck large stream running N. & soon after saw columns of steam, assuring us that we were on Fire Hole river." They arrived just as a large geyser was erupting, and had the good fortune to observe six major geysers in action in a brief stay of only twenty-four hours in the Upper Geyser Basin.

Langford wrote of their geyser watching:

> We gave such names to those of the geysers which we saw in action as we think will best illustrate their peculiarities. The one I

> have just described General Washburn has named 'Old Faithful,' because of the regularity of its eruptions, the intervals between which being from sixty to sixty-five minutes."[19]

Because it is out of Henry D. Washburn's inspiration that we have that marvelously apt place name — Old Faithful — it is only proper that he should speak for the party.

> This is the most remarkable valley we found. Hot springs are almost innumerable. Geysers were spouting in such size and number as to startle all, and are beyond description. Enormous columns of hot water and steam were thrown into the air with a velocity and noise truly amazing. We classified and named some of them according to size.
> No. 1. The Giant, seven by ten feet, throwing a solid column of water from eighty to one hundred and twenty feet high.
> No. 2. The Giantess, twenty by thirty, throwing a solid column and jets from one hundred fifty to two hundred feet high.
> No. 3. Old Faithful, seven by eight, irregular in shape, a solid column each hour seventy-five feet high.
> No. 4. Bee Hive, twenty-four by fifteen inches, a stream measured two hundred and nineteen feet.
> No. 5. Fan Tail, irregular shape, throwing a double stream sixty feet high.
> No. 6. is a beautiful arched spray, called by us the Grotto, with several apertures, through which, when quiet, one can easily pass, but when in action, each making so many vents for the water and steam.[20]

Unlike most of the notable place names considered thus far, Old Faithful has not proliferated within Yellowstone Park. The principal variant is the name of Old Faithful Inn, given in 1904 upon completion of that grand structure. There had been several "Old Faithful camps" — tent camps, that is — and an earlier hotel under that name, though it was familiarly called "The Shack Hotel."

The Old Faithful name was not regionalized within the Park, but remained the associative of the geyser and the inn. It has not fared so well outside the boundaries, where it has been used to designate many things — from a Pullman car to a brand of beer! Perhaps these are only evidence of its popularity.

## SHERIDAN

The name of General Philip H. Sheridan, the dashing Union cavalry officer of the Civil War, was given to the highest point of the Red Mountains in 1871, and that effectually ended the "colorful" confusion that had engulfed this isolated mountain mass.

On September 7, 1870, Langford and Doane climbed a summit east of the estuary of the upper Yellowstone River — the one now known as Colter Peak — for the purpose of viewing the country south of Yellowstone Lake. From that lofty place they were able to make out the terrain in a southwesterly direction as far as "a high mountain of a yellow rock, forming a divide or watershed in the center of the great basin, beyond which the waters flowed south and west." Five days later, during the search for Truman C. Everts — who had dropped behind the expedition and was separated from it on September 9 — Doane referred to the same height as "Yellow Mountain."[21]

Langford, who reached a point closer to the mountain during the search, saw it instead as "an immense brown mountain," which he described thus: "At the foot of Brown Mountain is a lake about a mile in width, the outlet of which on the south side is the source of Snake River, or the Lewis Fork of the Columbia."[22] The capitalization of "Brown Mountain" is a hint that he, like Doane with his "Yellow Mountain," also was thinking in terms of a place name.

But the Hayden Survey of the following year (1871) saw the mountain in yet another light, for it appeared on the map Dr. Hayden published that fall as "Red Mt." (see Note 6). He obviously did not consult with captains Barlow and Heap (whose party was traveling in tandem with Hayden's in order to share the same military escort), for they had an entirely different idea as to what the mountain's name ought to be.

Due to another "lost man" incident — or rather, two lost men — Barlow camped at the Heart Lake Geyser Basin and had a reason to climb to the summit on August 11. The description in his report is concluded with, "This mountain I have named Mount Sheridan."[23] It was so called in honor of General Philip H. Sheridan, whose great interest in the upper Yellowstone country led to the sending of army engineers (captains Barlow and Heap) to explore it that summer.

That was naming enough for one mountain, and it has stuck — but it did not satisfy everybody. Truman C. Everts, the member of the Washburn party who was lost in that wilderness for thirty-seven days, spent much of the time in a drafty bower among the hot springs at the foot of the mountain. He thought *he* had earned the honor of its naming: "I recognized the mountain which overshadowed it [his camp] as the landmark which a few days before had received from Gen. Washburn the name of Mount Everts . . . and as it is associated with some of the most agreeable and terrible incidents of my exile, I feel that I have more than a mere discoverer's right to the perpetuity of that christening."[24]

Everts is mistaken in his statement that the mountain he sheltered near had been named "Mount Everts"; there is no support for that in the writings of other members of the party (although it was so shown on the General Land Office map prepared by Washburn's successor, Surveyor-General Blaine, in 1871). On the other hand, both the diary of Cornelius Hedges and the "Mount Everts" article written for a Helena, Montana, newspaper soon after the return of the expedition (and before the rescue of Everts), leave no doubt which height was to be his "fitting" memorial.

Hedges and Everts left the camp on the shore of the Southeast Arm of Yellowstone Lake (southwest of the Molly Islands) after breakfast on September 8, to climb "a high mountain at S. end of Lake & on left of trail, called in honor of Everts. . . . We got back to camp before the train started [and] moved out in rear with Langford."[25] The climb was made up the north end of Two Ocean Plateau, which abuts on the Southeast Arm. It is not really a mountain, but only the ridge paralleling the upper course of the Yellowstone River on the west, and its northern end is not sufficiently distinct to merit a name on current maps. That the party concurred in Hedges's proposal is clearly stated in the "Mount Everts" article: "The company, of course, assented to my proposed name for the mountain we had visited, and let future tourists respect this monumental record."[26]

Neither Everts's companions nor those "future tourists" were content to let his name remain on the original Mount Everts. It was moved to its present location — on the escarpment east of Gardner River, opposite Mammoth Hot Springs — by mistake. And not by a simple mistake, but by mistakes compounded.

It appears that General Washburn prepared a manuscript map, either during the course of the expedition or upon his return to Helena, and it shows a "Mount Everts" vaguely entered between "Tower Cr." and "Gardner's River."[27] This would seem to represent a decision, soon after the rescue of Everts, to put his name on an eminence in the vicinity where he was found. It was a legacy that resulted in several efforts to honor him with a place name in the northern part of the Park.

The first effort was a by-product of the Barlow-Heap party's exploration in 1871. Although they did not show a Mount Everts on their official map, the name was evidently known from the General Land Office map file or from local informants, for Captain Barlow later mistakenly identified it with the volcanic summit now known as Bunsen Peak:

> *Mount Everts:* During our stay at this camp [Mammoth Hot Springs] I made a circuit of the high red-topped mountain to the west of the springs, near which Mr. Everts, who was lost in the Yellowstone basin the previous summer, was picked up by the men who went in search of him. The mountain stands in the forks of the branch of Gardiner's river, having cañons on two sides, through which the waters of the two forks flow in beautiful cascades, with one or two considerable falls.[28]

The Hayden Survey did no better in 1872 — possibly from listening to the garbled stories of locally employed packers and hunters. Somehow, the story of Everts's rescue and a "lost trail" incident that occurred earlier were confused, resulting in a serious historical warpage.

Upon entering what is now Yellowstone Park, the Washburn party passed to the east of Mount Everts into the Blacktail Deer Creek drainage. They reached that area by a small stream the advance party called "Antelope Creek" (for the animal killed in the pass at its head), but known to Sam Hauser as "lost Trail Creek."[29] Nowhere in the writings of the explorers of 1870 is there any reference to a "Rescue Creek," or any intimation that Everts was rescued on the stream they called "Antelope" and "Lost Trail."

On the other hand, the place where Truman C. Everts was found is known beyond question. It was not on a stream draining from Mount Everts — or on any stream — but on the open slope of Crescent Hill, facing The Cut. The authority for this is the man who found Everts — "Yellowstone Jack" Baronett. In the summer of 1886, a writer by the name of Theodore Gerrish interviewed Baronett concerning the rescue of Everts, and that interesting account ended with this statement: "The mountain which bears his name is six miles from the place where I found him."[30]

Historian Chittenden was able to be more specific in a letter to Captain George Anderson in 1895:

> By the way I got an interesting letter from Barronett [*sic*] lately. It was he, you know, who found Everts. I was surprised to learn that he found him over near the 'Devil's Cut' (or 'Gut') and not on Mount Everts at all. He says he piled up a mound of stones where he found him. I find this statement confirmed by Norris who visited the spot when he was Supt."[31]

Returning to the Sheridan name, there have been a number of spin-offs. Two, "Sheridan Geyser" (1881) and "Sheridan Trail" (1882), have not survived, as mentioned in Chapter 6. Another use of the general's name that has not survived is "Camp Sheridan," the designation given the cantonment built in September 1886 (at the foot of the main terraces at Mammoth Hot Springs) as housing for the detachment of United States Cavalry that took over administration of the Park August 20, 1886. Camp Sheridan was headquarters for the troops stationed in the Park until permanent facilities were begun at the "new post" — Fort Yellowstone — authorized May 11, 1891. As the fort took shape on the eastern edge of the "Hotel Terrace," the Camp Sheridan buildings were relegated to less important uses until 1915, when most of the original buildings were razed and the site lost its significance.

Another Sheridan name was added to the Park topographic map in 1956, when the Geological Survey put the general's name on a small lake at the foot of Mount Sheridan, on its southeast side.

## LEHARDY

While the usage associated with this name lacks the broad significance of those already discussed, it does have an interesting origin. It is also a reminder of the belated Jones expedition.

Until 1957 the rapids in Yellowstone River 2.9 miles north of Lake Junction were not shown on the Park map (though they were known unofficially as the "Yellowstone Cascades" or "Yellowstone River Rapids"). In preparing the fifty-eighth edition of the *Haynes Guide*, Jack Haynes added on page 126, "LeHardys Rapids (named 1873, Capt. Jones Rept.)," and thereby resurrected a place name of historic interest.

LeHardys Rapids. The name given this foaming barrier to navigation of the Yellowstone River remembers the "shipwreck" of two topographers who attempted to float down the river on a raft in 1873.

An information panel soon appeared at the parking area above the rapids to explain the origin of the name: "These rapids brought near tragedy to members of Captain Wm. A. Jones' exploration party. It was here the raft used for mapping the river's course capsized on August 7, 1873. This narrow escape ended the project and Captain Jones named the rapids for LeHardy, one of the men involved."

That brief statement, presenting nearly all that was known of the incident, came to the attention of a relative — Mrs. J. B. Gauntt of Chattanooga, Tennesee — who contacted the Park superintendent's office with information on the whereabouts of Paul LeHardy's son. As a result, the Park obtained a copy of the diary LeHardy kept while traveling through the area in 1873. According to his account, LeHardy and another topographer named Gabbet made a raft while the party was camped at the outlet of Yellowstone Lake, intending to go out to Stevenson Island. The wind prevented such a voyage, so they decided to float down the Yellowstone, mapping its course as they went, until they could rejoin the main party above the Yellowstone Falls.

All went well for a time, their raft — loaded with bedding, food, and guns — drifting serenely along on clear water that was alive with large trout.

> After a while we saw that a steep bluff would cause a sharp bend in the stream, indeed the water on the inside of the curve which we hugged because it was the shortest way, became shallow and quite rapid as we approached the bend. Realizing that there would be more water for our clumsy, unwieldy raft on the other side, we tried, but in vain, to cross over, now we perceived that we were fast approaching a fall or rapids. It was now too late for anything except to trust to luck. On the crest of the rapids there appeared an opening we could shoot through, but upon reaching this gap we perceived, half way down the chute, a conical rock and instantly our heavy raft straddled it, the rear end being at once sucked to the bottom. The water fortunately was only about two feet deep. We quickly pushed our paraphernalia up out of reach of the water and then considered what to do next. As we looked at the galloping waters at the foot of the bluff and wishing we had gone down on that side, we saw our outfit filing past up above and looking down at us. They could not have done anything to help us. Gabbet stripped, I tied a rope about his waist and he let himself down into the eddy below the rock then worked his way to the stony shore. I had tied the rope to the submerged side of the raft and the idea was that when we had transfered our stuff to the shore I would go over and we two would pull to dislodge the raft if possible. The rope broke and there we were marooned on the wrong side of the river. We 'cached' much of our surplus, including my precious LeFalcheux shotgun, which got broken when thrown ashore. Then we followed a game trail down the river.[32]

The two topographers reached the main party after more adventures of no particular importance here. The Geological Survey recognized the LeHardy Rapids name on the 1959 Canyon Village (Wyoming) quadrangle, thus legitimizing a usage already 82 years old.[33]

## MISCELLANY

The period of definitive exploration left more than two hundred place names, many of which have no stories of particular interest behind them. However, some of that nondescript naming had results worth mentioning.

The Raynolds Expedition of 1859–1860 contributed seven place names related to what is now Yellowstone Park, and five of those came from guide Jim Bridger. Thus, they are essentially of fur trade origin and are included in Chapter 2. One of the exceptions was "Mt. Gallatin," a doubtful case at best. The other was the sobriquet, "Valley of the Upper Yellowstone" — that awe-inspiring and mysterious name that set the stage for the later "Wonderland." It also furnished the Folsom party (1869) with just the right title for the magazine article that publicized their discoveries in that inscrutable area.[34]

The earlier statement that the Folsom party's influence on Park nomenclature was slight is true; few place names can be attributed to them. Cartographically, however, the expedition achieved significance through its correction of a long-standing misconception in the shape of Yellowstone Lake. All prior maps had portrayed that body of water as essentially spindle shaped (as shown on the Lewis and Clark map of 1814), but Folsom gave it an outline that is recognizably correct. The Folsom party was also the first to differentiate the "Great Falls" of Yellowstone River as Lower Falls and Upper Falls — citing their heights as 350 and 115 feet, respectively, which compares well with the accepted figures of 308 and 109. A place name — Overhanging Cliff — might be theirs, but that depends on whether they were naming or only describing when they wrote, "We picked our way to a timbered point about mid-way of the cañon [The Narrows], and found ourselves upon the verge of an overhanging cliff at least seven hundred feet in height."[35] At any rate, that *is* the name now, and no one else seems to have as good a claim. Obsolete place names used by the Folsom party include "The Chemical Works" — the present Hot Spring Basin Group — and "Valley of Death" for the Firehole Geyser Basins (which echoed a concern of prospectors with regard to an apparent absence of living things). The two maps that resulted indirectly from this exploration carried some place names that were "new" in the sense that they had not been shown previously, even though of undoubted prospector origin.[36] These are: "Third Canyon" (now Black and Grand Canyons), "East Fork" (now Lamar River), "Burning Spring" (now Calcite Springs), and a badly placed "Alum Creek," which would be a source of much misunderstanding.

This is a good place to mention a feature the Folsom party did *not* name, though they described it well enough. During their layover in the Tower Fall area, Cook and Folsom climbed to a vantage point from which they could see nearly the whole distance through The Narrows, and made the following observation:

> A short distance above us, rising from the bed of the river, stood a monument, or pyramid, of conglomerate circular in form, which we estimated to be forty feet in diameter at the base, three hundred feet high, diminishing in size in a true taper, to its top which was not more than three feet across. It was so slender it seemed as if one man could topple it over. How it was formed, I leave to others to conjecture."[37]

Lt. Gustavus C. Doane of the Washburn party (1870) saw this spire standing on the margin of the river and called it "Column Rock, a huge pile of alternate layers of basalt and amygdaloid cement, several hundred feet in height, surmounted by a pinnacle of trap." His naming of this unusual feature seems to have been overlooked, possibly because Langford used the same name in describing the long rampart of columnar basalt that caps the east wall of The Narrows. Neither name stuck, but with the building of the road from Canyon over Dunraven Pass to Tower Fall in 1903, this feature came to the attention of engineer Hiram M. Chittenden. Recalling from his familiarity with the Park's history that the Folsom-Cook explorers had described the column in 1869, he suggested to Arnold Hague that it ought to be named "Folsom's Column," but Hague demurred on the grounds that the Park had enough personal names attached to its features.[38]

As a result, nothing official was done and Column Rock still lacks an approved name, though a number of unofficial usages have been associated with it over the years. Probably the best were "Cleopatra's Needle" and simply "The Needle," both of which appeared in 1903. The former never really caught on, while there was a problem with the latter in that the Park already *had* a feature by that name. The Geological Survey had named a spire on the divide between South Cache Creek and the main branch of that stream "The Needle" in 1885, and it would have been confusing to duplicate that approved usage. In view of this, Doane's "Column Rock" ought to be confirmed.

The Washburn party applied names with a satanic connotation to several features they discovered, and in so doing set a pattern enthusiastically followed by later visitors. Their use of "Devil's Slide," "Hell Broth Spring," "Devil's Hoof," and "Devil's Den" suited a traveling public raised on a "hell-fire and brimstone" theology, even though Langford had obvious misgivings. When he used "The Devil's Den" as an apt descriptive term for the tangle of rocks and boulders along Cascade Creek above Crystal Falls, he added this:

> I presume that many persons will question the taste evinced by our company in the selection of names for the various objects of interest we have thus far met with; but they are all so different from any of Nature's works that we have ever seen or heard of, so entirely out of range of human experience, and withal so full of exhibitions which can suggest no other fancy than that which our good grandmothers have painted on our boyish imaginations as a destined future abode, that we are likely, almost involuntarily, to pursue the system with which we have commenced, to the end of our journey. A similar imagination has possessed travelers and visitors to other volcanic regions.[39]

The Devil's Den name did not remain long where the Washburn party assigned it. The Hayden Survey of the following year shifted it to the Tower Fall area, where the name has survived to the present as an adequate description of the course of Tower Creek immediately above its fall. In this designation, there is reason to suspect the men of the Hayden Survey party of 1871 were troubled by their own gloomy memories of that bloody corner of the Civil War battlefield of Gettysburg where the name came to prominence eight years earlier.

None of the satanic place names given to features within the present Park by the Washburn party survived. Their "Hell Broth Spring" is now Washburn Hot Springs (named by the Geological Survey in 1904 for General Henry D. Washburn), and the "Devil's Hoof," a peculiarly shaped spire of conglomerate rock above Tower Fall, now goes unnamed. The Devils Slide on Cinnabar Mountain two miles north of the Park boundary is the only remnant of their satanic nomenclature in current use.

It must be added that the trend toward satanic place names for Park features, so popular until World War I, has since fallen into disfavor. For a time, the devil had a proprietary interest in thirty-nine Park features, with the word "hell" appearing in eight others and "satan" in one. But in the last half century he has been largely dispossessed, so that his influence is now minimal. A mere three place names remain: Devils Thumb and Devils Kitchen at Mammoth Hot Springs, in addition to the Devils Den already mentioned (unless one includes the "Hellroaring" names remaining from the era of the prospectors, in which case there are six). More will be said about satanic nomenclature in Chapter 8.

What has been said thus far covers the great place names of this period, yet it leaves many unaccounted for. The three years of definitive exploration with which this period ended brought to light a large number of features that needed names in order that they might be described, and the need was supplied in proportion to the activity — (1869, six; 1870, fifty-eight; and 1871, a hundred thirty-five). This resulted in much confusion and many inappropriate usages.

Something has already been said of the confusion that surrounded the Mount Everts name and its companion, Rescue Creek, but that does not cover a much more common confusion resulting from separate parties using the same name for different features. Bad as that was, the men of the Hayden Survey added a special twist by occasionally using several entirely different names for the *same* feature.

Alum Creek is a good example to begin this discussion. The name was a legacy of the prospectors, probably referring originally to some stream draining the Mirror Plateau, but exactly which one seems beyond proof.[40] Lieutenant Doane made a good start by showing an "Alum Cr." entering the Yellowstone River nearly opposite Mount Washburn; then he placed another on the west side of the river above Cascade Creek (this is the Alum Creek of present maps). General Washburn confounded matters further by placing the "Alum Creek" name on the stream the prospectors had already named Pelican Creek, and a year later the Hayden Survey tried to make an "Alum Creek" out of the small stream now known as Alluvium Creek.

The "Antelope Creek" mentioned as a Washburn party name for present Rescue Creek was also intended to include the entire Blacktail Deer Creek drainage, but its failure to take root there left the name open for further use. In 1878 Superintendent Norris applied it to the stream draining the northern flank of Mount Washburn — a usage that has been accepted ever since. Yet there have been two attempts to use the Antelope Creek name elsewhere: in 1882 it was applied to present Trout Creek, and in 1933 there was an attempt to place it on present Elk Antler Creek.

The "Bee Hive Geyser" of the Washburn party fell victim to an attempt at robbery of a sort when the Hayden Survey named another thermal feature in the Lower Geyser Basin "Bee Hive," and then put the same name on a cone at Mammoth Hot Springs (this latter feature is now the Devils Thumb; the other can no longer be identified). The original feature has retained its name in slightly altered form as Beehive Geyser.

Comet Geyser is another place name with a confused past. The name originated with the Barlow party (1871) and was shown on the Hayden Survey map of 1878 as a central feature of the Daisy Group. However, that does not agree with Barlow's statement: "found the camp located on the east side of the river, near a small pine grove [in the vicinity of the Grand Geyser] . . . the fine geyser near my camp, named by me the 'Comet,' was entirely unnoticed by the party visiting the valley last summer."[41]

That the geyser seen and named by Barlow may not have been the one mapped finds some support in Riley's 1889 guidebook, where he says of the Oblong Geyser: "This geyser has sometimes been called the Comet; but the name given it in 1872 is more appropriate."[42] The Comet name has been applied to both the Daisy and the Splendid geysers.

Among the many place names given to Park features by the Hayden Survey in 1871 are some blunders that are more or less the natural result of work done in haste and published without adequate checking. This less-than-professional tendency of the Geological Survey of the Territories (Hayden Survey) undoubtedly resulted partly from the need to show accomplishment in order to obtain funding, and partly from employment of assistants with less-than-adequate training (eager young fellows related to important people).

The problems created, cartographically speaking, go beyond the misplacing of Mounts Langford and Doane and the inappropriate location of Mount Everts and Rescue Creek, as already described. A "Two Ocean Pass" was also placed between Grouse and Outlet creeks, where the drainages are not interconnected as at the real pass of that name south of the Park.

A tendency of both 1871 parties, and one that would not be approved under the present rules governing the naming of features, was a practice of naming things for persons they felt indebted to, for each other, and for family members. In addition to Mount Sheridan, already discussed, the Barlow party named Mount Hancock for General Winfield S. Hancock, commanding the army's Department of Dakota, and Mount Humphreys for Andrew A. Humphreys, the army's chief of engineers and the boss of captains Barlow and Heap.

The men of the Hayden Survey were more personal in the names they bestowed. Dr. Hayden named two features — a mountain and an island in Yellowstone Lake — for assistant James Stevenson, explaining his action as follows:

> In honor of his great services not only during the past season, but for over twelve years of unremitting toil as my assistant, often times without pecuniary reward, and with little of the scientific recognition that usually comes to the original explorer, I have desired that one of the principal islands of the lake and one of the noble peaks reflected in its clear waters should bear his name forever."[43]

In order to complete the record, it is necessary to add a statement that appeared in the year of Stevenson's death: "Against his protest the largest island in the wonderful Yellowstone Lake bears his name, and the loftiest peak that overlooks it is called 'Mount Stevenson.'"[44]

Names given for less worthy members of the Hayden Survey were Mount Chittenden for topographer George B. Chittenden (no relation to engineer-historian Hiram M. Chittenden of a later time) and Carrington Island for zoologist E. Campbell Carrington (such a mere speck of ground that one senses a joke behind the naming, but if so, it was not recorded). Two personal relationships appear — Frank Island for Frank Elliott, the brother of artist Henry W. Elliott, and Mary Bay (at first, "Mary's Bay") for Mary Force, the sweetheart Elliott did *not* marry. Another lady's name, appearing as Mollie Island on the Barlow map and an apparent relic of that party, is harder to explain. For years the statement of Historian Chittenden — "For the wife of Mr. Henry Gannett" — was accepted; but astronomer Gannett did not join the Hayden Survey until the following year, so it must have been someone else's "Mollie" who is honored by the Molly Islands name.

A place name that originated with the first Hayden Survey party, though it now persists in a different place, is also of interest here. Former Chief Park Naturalist C. Max Bauer credits photographer W. H. Jackson, whom he knew well, with identifying present Grand View as the "Artist's Point" from which Thomas Moran sketched the details for his huge oil painting, "The Grand Canyon of the Yellowstone."[45] However, the Geological Survey applied that name to another point on the other side of the Grand Canyon in 1883.[46] The name persisted at that improper location, probably because of a mistaken impression that Moran had worked there. The United States Geographic Board made that mistake official in its *Decisions* of May 7, 1930. Yet a lingering tendency to use the name as Hayden had led to further clarification in the *Decisions* of June 30, 1937:

> *Lookout Point:* a point on the north rim of Grand Canyon, Yellowstone National Park, Wyo. Named by common usage about 1900. Prior to that time, Lookout Point and Grand View Point were known by various names, the Hayden Survey calling one Morans Point and the other Artist Point, but because there was some confusion as to which name referred to Lookout Point, the older name was dropped. (pp. 18 & 19)

In that manner the Artist Point name was confirmed where it didn't really belong.

Several names from the twilight of the exploratory period — represented by the Jones party's reconnaissance in 1873 — will do to close this chapter. Captain William A. Jones led his party into the Yellowstone region by way of the North Fork of Shoshone River and what is now called Jones Pass. A consequence of that crossing was noted in Chapter 1, where the captain's renaming of the "Yellowstone Mountains" as the Absaroka Range was discussed. A further result was the naming of both the pass (by Superintendent Norris in 1880) and the stream by which he reached it (by the Geological Survey in 1893) for Jones, by then a major.

Captain Jones followed the right or east rim of the Grand Canyon below the falls of the Yellowstone, naming Deep Creek as he passed that stream. In the Lower Geyser Basin, Theodore B. Comstock, the expedition's geologist, named Clepsydra Geyser, thus comparing it to the water clock of the ancient Greeks. Comstock also named two geysers in the upper basin: "Locomotive Spring" (now Locomotive Vent) and "Impulsive Geyser," a feature already named Jet Geyser by the Hayden Survey.

While passing around the south shore of Yellowstone Lake, the Jones party named The Promontory. But the expedition's most important contribution, cartographically speaking, was its determination of the true nature of Two Ocean Pass. Leaving the Park by way of the old trapper's trail up the valley of the upper Yellowstone River, these explorers were able to verify the existence of that geographical curiosity Jim Bridger called the "Two Ocean Water." However, it did not get its present name — Two Ocean Pass — until it was mapped by the Geological Survey in 1911.

## NOTES

1. R. G. Thwaites, ed., *Original Journals of Lewis & Clark, 1804–1806* (New York: Dodd, Mead & Co., 1904–1905), II, p. 281.
2. William Clark, "A Map of Lewis and Clark's Track Across the Western Portion of North America, from the Mississippi to the Pacific Ocean," Copied by Samuel Lewis, 1806. National Archives, RG–77, Office of Chief of Engineers, US 529. This map was published in Nicholas Biddle's *History of the Expedition Under the Command of Captains Lewis and Clark* (Philadelphia, 1814).
3. William Raynolds, *Map of the Yellowstone and Missouri Rivers and Their Tributaries*, MS map compiled in 1860, National Archives, RG–77, Office of the Chief of Engineers, Q 106–1.
4. Ibid.

5 Walter W. deLacy, *Map of the Territory of Montana, With Portions of the Adjoining Territories . . . 1870* (New York: Colton & Co., 1870). Copy obtained from Library of Congress, Map Division.

6. Ferdinand V. Hayden, *Yellowstone National Park, 1871*. National Archives, RG–57, Hayden Survey, YNP.
7. *United States Statutes at Large*, Vol. 17, Chap. 24, Sec. 1, p. 32.
8. F. V. Hayden, *Sixth Annual Report of the United States Geological Survey of the Territories . . . 1872* (Washington: Government Printing Office, 1873), p. 243.
9. Philetus W. Norris, *Fifth Annual Report of the Superintendent of the Yellowstone National Park*, December 1, 1881 (Washington: Government Printing Office, 1881), p. 44.
10. Samuel T. Hauser, MS diary, Aug. 17 to Sept. 4, 1870. Coe Collection Item No. 249, Beinecke Library, Yale University, New Haven, CT.
11. Nathaniel P. Langford, *Diary of the Washburn Expedition to the Yellowstone and Firehole Rivers in the Year 1870* (St. Paul: F. J. Haynes, 1905), pp. 21–22. It must be noted here that this is not truly a "diary," but only an account back-written from several sources. No Langford diary covering the period of the expedition is known to exist today.
12. Cornelius Hedges, *Diary of Cornelius Hedges from June 24, 1870 to October 16, 1871*, No. 2165, Montana Historical Society, Helena. Entry for Aug. 28, p. 21.
13. Warren C. Gillette, *Diary of Warren Caleb Gillette from August 17 to September 27, 1870*, MS diary, Montana Historical Society, Helena. Entry for Aug. 29, p. 22.
14. Cornelius Hedges, "Address delivered by Cornelius Hedges on Sunday evening January 29th, 1871, eulogistic of Hon. Henry D. Washburn, at the M.E. Church, Helena, MT." Hedges Papers, Montana Historical Society, Helena.
15. Hedges, *Diary*, p. 27; see Note 12 for citation.
16. Philetus W. Norris, *Report Upon the Yellowstone National Park, to the Secretary of the Interior, for the Year 1878* (Washington: Government Printing Office, 1879), map.
17. Langford, *Diary*, p. 41; see Note 11 for citation.
18. Hedges, *Diary* (entry for Sept. 18, p. 37); see Note 12 for citation.
19. Hedges, op. cit., p. 108.
20. Henry D. Washburn, "The Yellowstone Expedition," *Helena* (MT) *Daily Herald*, Sept. 28, 1870.
21. Gustavus C. Doane, *Official Report of the Washburn-Langford-Doane Expedition into the Upper Yellowstone in 1870*, a hand-written copy made at Fort Ellis, Dec. 15, 1870. Montana State University Library, MS 492, pp. 40 & 44.

22. Nathaniel P. Langford, "Notes of a Lecture Given by N. P. Langford During the Winter of 1870–1871," handwritten in ink in a notebook. MS, Yellowstone Park Reference Library). See pp. 145 & 147.
23. John W. Barlow and David P. Heap, *Report of a Reconnaissance of the Basin of the Upper Yellowstone in 1871* (Washington: Government Printing Office, 1872), p. 34.
24. Truman C. Everts, "Thirty-seven Days of Peril," *Scribner's Monthly* 3/1 (Nov. 1871), p. 4.
25. Hedges, *Diary* (entry for Sept. 8, p. 30); see Note 12 for citation.
26. Cornelius Hedges, "Mount Everts," *Helena* (MT) *Daily Herald*, Oct. 8, 1870.
27. Henry D. Washburn, *Route of the Washburn Party, 1870, being a tracing of map made to accompany Washburn's report to Dept. of Interior.* National Archives, RG–49 (OMF), Records of the Gen. Land Office, Mont. 5.
28. John W. Barlow, "Natural Wonders," *Chicago Evening Journal*, January 13, 1872.
29. Hauser, op. cit., entry for Aug. 26. On the same date, Gillette noted "camp tonight on a stream in the mountains about 6 miles from last camp Doane E [Everts] & soldier ahead could not find their trail"; Hedges added, "lost our trail & wandered about a good deal."
30. Theodore Gerrish, *Life in the World's Wonderland* (Biddeford, Maine: n.p., 1887), p. 240.
31. Hiram M. Chittenden, letter to Acting Superintendent George S. Anderson, Yellowstone National Park, March 28, 1895. Original in Yellowstone Archives.
32. Paul LeHardy, Sr., *An Autobiography of Paul LeHardy,* covering experiences while a member of the Jones Expedition into the Yellowstone country in 1873. Yellowstone Park Reference Library, p. 99.
33. See *Map of a Military Reconnoissance, in North Western Wyoming,* made under direction of Brig. Gen. E. O. C. Ord, Commanding the Dept. of the Platte, by Capt. W. A. Jones, 1873 [Published by the Chief of Engineers, 1874].
34. Charles W. Cook, "Valley of the Upper Yellowstone," *Western Monthly*, July 1870, pp. 60–67.
35. Ibid., p. 62.
36. These maps are the 1870 deLacy, mentioned in Note 5, and the manuscript map submitted to the General Land Office by Henry D. Washburn, November 1, 1869. National Archives, RG 49, GLO (OMF) — Montana 3. Following his return from the Yellowstone in 1869, David E. Folsom was employed in the Helena office of the surveyor-general of Montana Territory, and so was able to add his valuable information to a later map forwarded from that office.
37. Charles W. Cook, David E. Folsom, and William Peterson, *The Valley of the Upper Yellowstone,* ed. by Aubrey L. Haines (Norman: University of Oklahoma Press, 1965), pp. 23–24.
38. Lee H. Whittlesey, *Yellowstone Place Names* (Helena: Montana Historical Society Press, 1988), p. 111.
39. Langford, *Diary*, p. 28; see Note 11 for citation.
40. The name appears first on the two General Land Office maps of 1869. On the one predating the visit of the Folsom party, it is on the longest branch of the stream where the Lamar River ought to be; on the later map, the one corrected by the Folsom party, the name seems to designate Deep Creek. The 1870 deLacy map (also revised to show the Folsom party's information) definitely associates the name with Broad Creek.

41. Barlow, *Report*, pp. 24 & 25; see Note 23 for citation.
42. W. C. Riley, *Grand Tour Guide to the Yellowstone National Park* (St. Paul, MN: W.C. Riley, Pub., 1889), p. 78.
43. Ferdinand V. Hayden, *Preliminary Report of the United States Geological Survey of Montana and Portions of Adjacent Territories, Being a Fifth Annual Report of Progress* (Washington: Government Printing Office, 1872), p. 5.
44. Director, United States Geological Survey, *Ninth Annual Report . . . for the Fiscal Year Ending June 30, 1888* (Washington: Government Printing Office, 1889), p. 43.
45. C. M. Bauer, *Place Names of Yellowstone National Park,* a typed MS, Yellowstone Park Reference Library, 1937, p. 2. Moran's eight-by-fifteen-foot oil painting, completed in 1872, was purchased by Congress for $10,000 and hung in the national Capitol; however, W. H. Jackson states that it was neglected, later badly restored, and that "he made a second one, a fine copy, still in the national Gallery."(*Time Exposure*, 1940, p. 200). Since his statement could be misunderstood, it must be added that I checked the holdings of the National Gallery in October 1970, and found that the original (1872) canvas was in storage in very bad condition. The curator's file contained a letter from the secretary of Interior to the Hon. S. Dillon Ripley, July 2, 1968, stating: "In 1950 two paintings by Thos. Moran, 'The Grand Canyon of the Yellowstone' . . . [were] given to this Department by the Congress." The large oil on exhibition in the Lincoln Room at that time was not a copy of the original, but an 1893 rendition from the watercolor sketch done in 1871 and sold to George D. Pratt, who donated it to the National Gallery in 1928. (Moran made and sold at least two more renditions of this scene; one is held by the Gilcrease Museum, Tulsa, Oklahoma, and the other is privately owned in Hawaii).
46. Director, United States Geological Survey, *Sixth Annual Report . . . for the Fiscal Year Ending June 30, 1885* (Washington: Government Printing Office, 1886), p. 104.

# 5

## FROM SURVEY NOTEBOOKS

| | |
|---|---|
| *Electric* | *Gibbon* |
| *Dunraven* | *Lamar* |
| *Baronett/Barronette* | *Riddle* |
| *Monument* | *Lone Star* |
| *Bechler* | *Miscellany* |

Some of the men who completed the definitive exploration of the Yellowstone wilderness in 1871 came back to the new Park the following summer to begin the detailed mapping of its vast area — an extent larger than the states of Rhode Island and Delaware combined. This physical inventory required accurate delineation of the mountains and valleys, the streams and lakes, *and* the thermal features that are the very *raison d'etre* behind the Park's existence. It has been a long and tedious undertaking, requiring the combined effort of several organizations and a century of time to produce an adequate map. A direct result of that mapping has been the accumulation of new place names, either through alteration or substitution where old usages were concerned, or through appropriate naming of previously unidentified features.

The serious mapping of Yellowstone National Park was begun by Dr. Ferdinand V. Hayden's Geological Survey of the Territories in 1872, but the work of his organization was interrupted following the 1878 season by the merger that combined it with several similar Western surveys to form the new United States Geological Survey (March 3, 1879). Cartographically speaking, the Yellowstone legacy of the Hayden Survey was a topographic map — at the scale of two miles to the inch and with a contour interval of 100 feet, covering approximately three-quarters of the Park — and large-scale maps of the important geyser and hot spring areas prepared as the base sheets for locating and identifying thermal features.

Mapping of the Yellowstone Park area, as resumed in 1883 by the United States Geological Survey, followed a nationwide plan under which work was done by "quadrangles" — unit areas that could be arranged as folios in a "geological atlas." The early sheets covered a full degree of latitude and longitude (at the scale of 1:250,000), so that the Park area was on two folios, *Livingston* and *Three Forks*. The introduction of thirty-minute quadrangles (at the scale of 1:125,000) in 1896 resulted in a *Yellowstone National Park Folio* of four sheets: Gallatin, Canyon, Shoshone, and Lake. Prior to World War II, a start was made toward producing a series of fifteen-minute quadrangles (at the scale of 1:62,500), but the project did not appreciably affect the Park area until the 1950s.

It might be helpful to explain that the earlier topographic mapping was done essentially as in the days of the Hayden Survey: by topographers sketching in the contours — lines of equal elevation — around observation points (usually mountain peaks) located by secondary triangulation within a nationwide primary triangulation scheme. The accuracy of such mapping depended in large measure on the skill of the topographer and the availability of satisfactory observing stations. A method of mapping by photogrammetry — by optical measurements made from aerial photographs — received such impetus during World War II that the Geological Survey changed to the new method during the postwar years.

The first fifteen-minute quadrangle involving a portion of the Park produced by the new method was the Crown Butte sheet (1955), and the work went forward so rapidly that twenty-five quadrangles were completed by 1960. That provided a basis for a new Park-wide map of consistently high accuracy, and also ended the complaints of several generations of rangers troubled by the inaccuracies of the older topographic maps.

The 1961 edition of the Park-wide map was not the first compilation covering the entire area. The need to combine the quadrangle sheets into a Park-wide map was evident very early, and just such a map was prepared from the thirty-minute sheets in 1904 (*Yellowstone National Park and Portions of Yellowstone Forest Reserve*). It was intended primarily for the use of the scouts and soldiers then responsible for patrolling both the Park and the adjacent forest reserve, but its small scale (1:250,000) limited the usefulness of the map. There was an improvement in 1912 with the printing of a topographic map of the entire area within the original boundaries of the Park at the better scale of 1:125,000. That administrative map was reprinted in 1928, and then revised extensively in 1930 to show the new boundaries authorized by the legislation signed into law on March 1, 1929 (changes concerned with putting much of the eastern boundary on the hydrostatic divide, and with adding an area of fossil forests to the Park's northwest corner). The 1930 map was reprinted in 1954, but without showing the "Gardiner Game Ranch" added to the Park by presidential proclamation on October 20, 1932.

That is a brief summary of the cartographic work done by the United States Geological Survey in Yellowstone Park over the eighty-two years prior to printing of the 1961 edition of the Park-wide map. During those years, the men of the Survey recorded many place names — more than have come to us from any other source — yet it is often difficult to tell when they were originators and when they were merely recording. For that reason, they may receive less credit than they are due. This is true also for other surveyors of past years — the men who ran the original boundary lines, the men of the U.S. Engineering Department, and the men who worked with U.S. Cadastral Engineer Bandy on the survey of the Park's east boundary in 1931.

## ELECTRIC

This most appropriate place name had its origin in the "electrifying" experience of a small Hayden Survey party during an ascent of the great peak that stands sentinel-like above the Northern Entrance to Yellowstone Park.

The party that ascended the peak on July 26, 1872, for the purpose of determining its height by angulation consisted of Dr. Ferdinand V. Hayden, astronomer Henry Gannett, and assistant topographer Alexander E. Brown. The following is Gannett's description of their experience:

> A thunder-shower was approaching as we neared the summit of the mountain. I was above the others of the party, and, when about fifty feet below the summit, the electric current began to pass through my body. At first I felt nothing, but heard a crackling noise, similar to a rapid discharge of sparks from a friction machine. Immediately after, I began to feel a tingling or prickling sensation in my head and the ends of my fingers, which, as well as the noise, increased rapidly, until, when I reached the top, the noise, which had not changed its character, was deafening, and my hair stood completely on end, while the tingling, pricking sensation was absolutely painful. Taking off my hat partially relieved it. I started down again, and met the others

Electric Peak (10,992 ft.). A party of surveyors had a "hair-raising" experience on the summit when caught in a summer lightning storm there in 1872.

> twenty-five or thirty feet below the summit. They were affected similarly, but in a less degree. One of them attempted to go to the top, but had proceeded but a few feet when he received quite a severe shock, which felled him as if he had stumbled. We then returned down the mountain about three hundred feet, and to this point we still heard and felt the electricity.[1]

Topographer Brown was fortunate he was not killed by the electrical discharge that knocked him down, and the party's situation remained dangerous even after they retreated from the summit. Dr. Hayden states that they waited until the storm passed over, and "about 4 o'clock in the afternoon we succeeded in reaching the top, and Mr. Gannett found the altitude of the peak to be 10,992 feet above the sea. We named it Electric Peak."[2]

It is interesting that the elevation determined for the summit by Henry Gannett at that time was not the elevation that appeared on the topographic map published in 1878. On that map Electric Peak is recorded at 11,155 feet, which was the accepted elevation for seventy-seven years — until 1955, when a resurvey put the mountain's height back at precisely 10,992 feet![3] Considering the legendary skill of astronomer Gannett, there is less reason for surprise in this latter-day verification of his original determination than in the real puzzle — why it wasn't used on the 1878 map.

The propriety of this place name — Electric Peak — has been questioned only once, as far as I am aware; and that was such a left-handed happening it must be explained in the words of one who was there. Jack Ellis Haynes was the son of photographer Frank J. Haynes, a major concessioner. While Jack speaks of himself and siblings as "youngsters," they were all adults, with Jack at 27 already involved in the family photographic business. Keep that in mind as he reminisces (the comments within brackets are mine):

> Along about 1911 Thomas Alva Edison, the eminent inventor, spent a lot of time at Mammoth Hotel and in other parts of the park, but he liked Mammoth best because they had a wonderful orchestra there and Mr. Edison was fascinated by the fine music [he was quite deaf but could hear the orchestra through a horn-like contrivance he held to his ear]. . . . Vice President Fairbanks and Alice Roosevelt Longworth and a lot of other dignitaries landed at Mammoth and Col. Lloyd Brett [acting military superintendent], of course, was all excited and decided to put on a big dinner in the dining room of the Mammoth Hotel in honor of Thomas Edison and those dignitaries . . . they got a little magnanimous and invited our family, my mother and father and us youngsters to the dinner also. So we went in to where this big banquet table was set. Col. Brett was the host and Mr. Edison sat across the table from him. When the dinner was practically over, Col. Brett rose to his feet and said: 'Ladies and Gentlemen, we have as one of our distinguished guests tonight, Mr. Thomas Alva Edison and you know his reputation for a knowledge of electricity is unequalled in any other individual in the world.' He made a lot of other complimentary remarks about Mr. Edison [a eulogy which Edison followed with his listening device]. . . . Then Col. Brett said: 'It seems to me that it is eminently fitting that the name of Electric Peak should be changed to Mount Edison in honor of this distinguished gentleman and I want to officially propose that the name of the highest mountain in the park be changed to Mount Edison.' Everybody applauded, with 'hear, hear,' and all that stuff. But I didn't applaud because I didn't like the idea; however, practically everybody there seemed to agree with Col. Brett. After the din had subsided a little, Mr.

> Edison laid aside the thing he had been holding up to his ear and said: 'Ladies and Gentlemen, I appreciate what has been said and all that, but,' he said, 'it would be my preference to leave the name Electric Peak and not change it for anybody, or any purpose. . . . I think that is a suitable name for that magnificent peak and I prefer that the name not be changed.' With that he sat down, everybody applauded, and Col. Brett hemmed and hawed a few moments then said that he would be glad to defer to Edison's wishes in the matter.[4]

Electric is a usage that has not proliferated greatly. The drainage out of the cirque on the north side of Electric Peak is Electric Creek, which runs into Reese Creek. On the west bank of Yellowstone River, where the ruined coke ovens north of Reese Creek are located, there was formerly a coal-mining town that was renamed Electric in 1904. The presumed reason for renaming, though not the *real* reason, was explained in a local joke drawn from the original name of the town. It was Horr, named for "Major" Harry R. Horr, who claimed and developed the coal measure found nearby.

Harry was something of a promoter who sold out to a company that styled itself the Electric Coke and Coal Company, and it was convenient to identify the company town with the company. That's the real reason, but local wags claimed the women of the town had insisted on the change "because they didn't care to live in Horr houses."

## GIBBON

An important headwater of the Madison River is identified with this place name, which was a product of military mischance.

Colonel John Gibbon, who commanded the scattered army detachments in Montana Territory, entered Yellowstone Park with Lieutenant Benson and a small escort during the first week of August, 1872. They came up the Yellowstone River, stopping briefly at Mammoth Hot Springs to enjoy the rude "spa" operated by Harry Horr and James McCartney. It is possible that they talked to Eugene S. Topping and Dwight Woodruff while there, or knew through camp gossip that those young men had discovered a new geyser basin (Norris) and a shorter route to the Firehole basins. At least that makes sense in view of what happened.

From Mammoth Hot Springs, Gibbon's party took the trail over the Washburn Range to the Grand Canyon (where they were seen by members of a Hayden Survey party on August 7). They then continued to the outlet of Yellowstone Lake and along the west shore to the hot springs at West Thumb, then proceeded due west through trackless forest, past Shoshone Lake, and on to the Upper Geyser Basin. How long they remained there and why they attempted a new, unproven route for the return to Mammoth Hot Springs is uncertain (they should have taken the trail up Nez Perce Creek to the Grand Canyon).

What is known is that they went down the Firehole River to Madison Junction and attempted to ascend the stream that enters from the east. According to Colonel Gibbon, they "fell short of provisions, and had to kill squirrels, blue-jays, and a pelican, and finally to grub for wild roots for subsistence."[5] They had to return to Madison Junction where Dr. Hayden's division of the Survey found them in a famished condition on August 20. The surveyors were also short of provisions — in fact they were on their way out of the Park by the Madison River route — but they shared what they had with the soldiers. Hayden afterward gave the stream that had baffled the soldiers the name of Gibbon's Fork, "in honor of General [a Civil War brevet] John Gibbon, United States Army, who has been in military command of Montana for some years, and has, on many occasions, rendered the survey most important services."[6]

Gibbon Falls. Col. John Gibbon's name was attached to this river and a number of features along it because of his failure to complete an exploration of the stream.

It must be added that Topping and

Woodruff conducted a party from Mammoth Hot Springs to the Firehole geyser basins that summer by their new route. That party included the first white woman to reach the Upper Geyser Basin — Mrs. H. H. Stone of Bozeman.

Hayden's usage — Gibbon's Fork — soon became Gibbon River, and that root has spawned many place names over the years. The first of these spin-offs was Gibbon Geyser Basin, used by the Hayden Survey to identify the large area of hot springs and geysers midway up the river. That same year — 1878 — Superintendent Norris used "Norris Geyser Plateau" to indicate the trail destination at the geyser basin.[7] During the two following years Norris substituted his own name for Gibbon's at the geyser basin and attempted to do the same for the upper reaches of Gibbon River, but with less success. This will be discussed at length in the next chapter.

John Gibbon was eased out of the principal geyser basin on his river, but he finally was identified with another basin. In 1939, the thermal area surrounding Gibbon Meadows (a name that originated with the U.S. Engineer Department in 1892) was named Gibbon Geyser Basin. It "is applied now to the scattered groups of geysers, hot springs, and other thermal phenomena beginning at the north edge of Gibbon Meadows and extending southward into Gibbon Canyon as far as the Beryl Spring area."[8]

Another Gibbon place name that lost out to Norris was Gibbon Lake at the head of Solfatara Creek. The Hayden Survey named it that in 1878, but Norris's poetic, though hardly original, appellation for that closely embowered body of water — Lake of the Woods — caught on with the public. Other Gibbon usages that have not survived are Gibbon Paint Pot Basin (1885); Gibbon Paint Pots (1896), now Artists Paint Pots; and Gibbon Chocolate Pots (1914), now simply Chocolate Pots.

But not all have been lost. Here are some survivors: Gibbon Falls, mentioned by H. B. Wiley in 1883 and probably a common usage; Gibbon Hill (1885); Gibbon Canyon, first mentioned in *Haynes Guide* (1896); and Gibbon Hill Geyser (1935). In all, there have been sixteen usages, half of which remain in use.

## DUNRAVEN

This place name is essentially a thank you for what a foreign visitor did for Yellowstone Park.

In 1874 a titled Englishman, the Fourth Earl of Dunraven at Adare, Ireland, visited the Park in connection with a hunting trip to the northern Rocky Mountains. In *The Great Divide*, a book covering his American experiences, 154 pages are devoted to Yellowstone Park. They are replete with charming vignettes of the new reserve and did much to interest Britons and other Europeans in it; but, even more important, he was able to give them a sense of its unique importance when he wrote:

> All honor then to the United States for having bequeathed as a free gift to man the beauties and curiosities of 'Wonderland.' It was an act worthy of a great nation, and she will have her reward in the praises of the present army of tourists, no less than in the thanks of the generations of them yet to come."[9]

Dunraven Peak (9,900 ft.). This summit and the pass between it and Mount Washburn was named for Lord Dunraven in thanks for his book, which introduced Europeans to our first national park.

In recognition of that service to the Park and the idea it represented, in 1878 Henry Gannett put Dunraven's name on the peak west of Mount Washburn in the Washburn Range. He says: "Southwest of Mount Washburn, and distant from it but 2 miles, is a sharp cone rising to a height of 9,988 feet. This I have named Dunraven Peak, in honor of the Earl of Dunraven, whose travels and writings have done so much towards making this region known to our cousins across the water."[10] Historian Chittenden has stated that P. W. Norris had already put his own name on that peak, but I have found nothing to support such a statement.

The year 1877 was the first as superintendent for Norris, and he did not reach the Park until July. The five weeks until his crippling injury near Tower Fall were all spent in the mountains along the northern verge seeking a feasible trail to the Yellowstone River by way of the Slough Creek–Rosebud drainages. It was not a season for naming things, as his annual report shows. His second season got off to another slow start, with his return to the Park on July 29. Then the Indian scare (Bannock War) and the consequent need to construct the "Norris Road" as a defensive measure prevented any other activity until September 26. By that time the good weather was gone and the Hayden Survey was on its way home — with Dunraven's name definitely in place on their map.[11]

Dunraven Pass, the only spin-off from the Dunraven Peak usage, appears to have originated with Norris in 1879,[12] and that is an indication he agreed with the name Gannett had given the peak.

## LAMAR

This usage replaced one of those "east forks" so common in frontier nomenclature, and always so unhandy in a place name.

The large tributary entering Yellowstone River from the east just above its third canyon (Black Canyon) was known to the fur trappers as "Beaver Creek" and to the prospectors as the "East Fork" of the Yellowstone River. It was that qualification, "of the Yellowstone," necessary to separate it from those other east forks (of Firehole, of Gardiner, of Madison, and of the Upper Yellowstone), that made the name either impossibly ponderous in a complete form or likely to be misunderstood when shortened.

The Hayden Survey accepted the river's name just as they got it from the prospectors — as "East Fork of the Yellowstone" — but the men of the Geological Survey had a better understanding of how such tributary streams should be named. They preferred a stream name to apply throughout the longest continuous reach of water in the drainage, with all tributaries identified by different names, thus avoiding the confusion inherent in the use of such crutches as "fork," "prong," or "branch."

Field work accomplished by parties of the Geological Survey in 1884–1885 resulted in the renaming of the East Fork of the Yellowstone in honor of L.Q.C. (Lucius Quinctius Cincinnatus) Lamar, a

Mississippian who served as the secretary of Interior from March 6, 1885, to January 16, 1888. Only a father who was a student of the classics and an admirer of the Romans could have burdened a son with a name so freighted with honor, but it was not given in vain; Lamar was an honorable man, and he immediately set about clearing up the scandal that had clouded Superintendent Robert E. Carpenter's regime.

Geologist Arnold Hague, then in charge of geography for the Survey, was called upon to assist the special agent investigating management practices in the Park. The action taken by the secretary undoubtedly met with his approval, for we have the statement of George L. Henderson that Hague was responsible for renaming the river,[13] and only such a feeling of approbation would explain the statement Hague made later:

> Other than the names bestowed in honor of the early explorers of this region, but few personal names have been placed upon the grander physical features. The most important name given is that in honor of the present Secretary of the Interior, who has shown great interest in the preservation and maintenance of the Yellowstone Park.[14]

This usage has its share of spin-offs: Lamar Canyon, also named in 1885 (where the Park's granitic basement rocks lie beautifully exposed); Little Lamar River (1904); Lamar Mountain, named by Cadastral Engineer Bandy in the course of the east boundary survey in 1932 (it cradles the river's extreme headwaters); and three administrative usages of recent origin — Lamar Ranger Station (the "Buffalo Ranch" of an earlier day), Upper Lamar River Patrol Station, and the Lamar River Trail (a pack route connecting the two stations).

There is one more place name that does not appear on Park maps, yet is in constant use. "Lamar Valley" is a common way of referring to the beautiful, open valley below the mouth of Soda Butte Creek. This is the "Secluded Valley" of trapper Osborne Russell.

## BARONETT/BARRONETTE

Among the pioneers of the Yellowstone region, none was more colorful than "Yellowstone Jack" Baronett, whose fitting memorial is the great peak that overshadows the Northeast Entrance road.

This quiet, capable Englishman was, in his adventurous lifetime, sailor, prospector, soldier of fortune, scout for the army, guide, an assistant superintendent of the Park, and builder of the first bridge over the Yellowstone River at any point. It was the latter — construction and operation of a toll bridge on the miner's road to Cooke City — that put him "on the map."

Although John H. Baronett (1829–1906) was probably his real name, he appears to have preferred Collins Jack, or simply Jack Baronett. He was the prospector commissioned by the friends of Truman C. Everts to find that unfortunate man after his loss from the 1870 Washburn party of explorers. Baronett did find the missing man (a barely alive scarecrow after thirty-seven days of wandering in the Yellowstone wilderness). But in the course of that rescue, he discovered something else: a particularly good place to bridge the Yellowstone River on the route to the new mining camp on Clarks Fork (in the vicinity of present Cooke City).

After delivering Everts to his friends, Baronett went back to the site just above the mouth of Lamar River and built his bridge over the Yellowstone during the low water of the 1870–1871 winter. His pack train bridge, for that is all it was at first, was just *the* bridge until 1877, when General William T. Sherman's inspection party first used the term "Baronett's Bridge."[15] The Hayden Survey map of the following year showed the bridge, but spelled Baronett's name with a terminal "e" (a common error). The Geological Survey corrected the

Baronett Peak (10,404 ft.). Named by the Hayden Survey in 1878 for John H. Baronett, the man who rescued the lost Truman C. Everts and built the first bridge over the Yellowstone River at any point.

spelling on its quadrangle sheets and on the 1904 Park-wide map, but with the abandonment of the old crossing in 1905, the name was not used on later maps.

The 1878 map of the Hayden Survey placed Baronett's name on the great peak in the northeast corner of the Park, where it remains, but as in the case of Baronett Bridge, it was misspelled by adding a terminal "e." The Geological Survey corrected that and all was well — until publication of the new Abiathar Peak fifteen-minute quadrangle in 1959 — and there was the Baronett name, misspelled again! This time it gained an extra "r" *and* an extra "e." Photographer Henry Bird Calfee used the name "Hermit Mountain" on his stereo pictures of Baronett Peak, but his usage did not take — fortunately.

## RIDDLE

This usage is a fine example of how incredibly persistent an error can be once it is well established.

Following the return of the Lewis and Clark Expedition, William Clark received geographical information from fur trappers operating in the northern Rocky Mountains, and those gleanings were entered upon a manuscript map he kept.[16] Among the features entered as a result of John Colter's winter trip in 1807–1808 were two lakes: "Eustis Lake," now Yellowstone Lake, and "L. Biddle," now Jackson Lake. But when Samuel Lewis made the engraving for the map that was published in the *History of the Expedition Under Command of Captains Lewis and Clark* (Philadelphia, 1814), Biddle became "Lake Riddle," and many later mapmakers helped to perpetuate the error. Among them were Arrowsmith (1814), Roberdeau (1818), and Brué (1834).

A map published by A. J. Johnson in 1865 was popular with the prospectors and miners drawn to Montana Territory by the gold rush of the 1860s. Still relying on Lewis and Clark cartography, it showed two place names in the general area of present Yellowstone Park: a sausage-shaped "Subletts or Yellowstone Lake," and "L. Riddle" close below it.[17] This may have helped to detach the Riddle name from the Jackson Hole area and leave it vaguely related to the Park plateau, as a passage in a letter written by Charles R. Sunderlee indicates: "Never have I seen so many wonderful and marvelous curiosities of nature, embraced in such a small extent of country, as those existing in the vicinity of Lake Riddle, Sulpher Mountain, and the Falls of the Yellowstone."[18]

The Hayden Survey's chief geologist in 1872 was Prof. Frank H. Bradley, who evidently knew nothing of the origin of the Riddle name but still had some definite ideas about it. He called Riddle Lake a fugitive name, "located at several places, but nowhere permanently. It is supposed to have been used originally to designate the mythical lake among the mountains, whence, according to the hunters, water flowed to both oceans."[19] Of course, that was only a garbled version of Jim Bridger's "Two Ocean River" story, which was verified the next year by the Jones Expedition (as Two Ocean Pass).

But Bradley thought the riddle ought to be solved by fixing the fugitive name to a Yellowstone feature, and topographer Rudolph Hering thought he had just the right place (a lake almost on the Continental Divide and two miles south of the West Thumb of Yellowstone Lake), "where the waters of the two oceans start so nearly together, and thus to solve the 'riddle' of the 'two-ocean-water.'"[20]

A surveyor's joke will do to finish this discussion of Riddle Lake. The outlet stream was mapped later — in 1885 — by a United States Geological Survey party. They called the stream Solution Creek, because a riddle should not be left without a solution!

## MONUMENT

From this word came a usage that cannot be traced to a survey notebook, and yet it was a product of the Park's boundary surveys.

There is a fine, clear stream that descends into the valley of the upper Yellowstone River out of an amphitheater of such magnificent mountains that the men of the Geological Survey could only call it Mountain Creek. That name was assigned in 1885 and appeared on all the maps of the Survey until 1961. Then, through a most unfortunate interpretation of earlier events, the name was changed to Monument Creek. What happened can best be explained by summarizing the survey of the old east boundary line of Yellowstone Park.

The Park's organic act specified that the east boundary line should be on the "meridian passing ten miles to the eastward of the most eastern point of Yellowstone Lake."[21] However, it was a long time before any of the boundaries were surveyed.

Establishment of the Yellowstone Park Timberland Reserve by executive orders of March 30 and September 10, 1891, led to provision of funds for the survey of the east and south boundaries (since those lines would be boundaries of the new reserve also). Accordingly, Assistant Commissioner Bowers of the General Land Office awarded a contract on June 4, 1892, to Philip Galligher to survey both lines for $8,960. Considering that the east line was sixty-one miles long and the south fifty-seven miles, the contract amount was probably less than adequate.

As it turned out, the work was also inadequate. The acting military superintendent, Captain George S. Anderson, complained about this fact in his 1894 report to the secretary of Interior:

> Mr. Gallagher [*sic*], who last year had the contract for running the boundary of the timber reserve, did not begin his work at the most easterly portion of the lake, nor did he make his line continuous from the point where it should have begun, to the north line of the Park, nor did he properly and conspicuously mark the line. If it should be found practicable to have an officer of the Corps of Engineers report to me for temporary duty, I could have the line carefully run.[22]

Captain Anderson was a line officer with a considerable knowledge of engineering. He had successfully ousted the Corps of Engineers from the Park and taken personal charge of road construction and maintenance activities, and that had enhanced his reputation. As a result, he obtained the services of an engineer officer — Captain Charles S. Bromwell.

Bromwell began his work from a stone monument placed on the shore of the Southeast Arm of Yellowstone Lake at the mouth of Beaverdam Creek. Finding the terrain between that point and the meridian of the east line impossibly rugged, he was forced to approach the east boundary in an indirect manner. He ran a traverse line down the valley of the upper Yellowstone River to Mountain Creek, and then up that stream to the meridian of the boundary, where he placed a marker. This zero-mile post was approximately 5.3 miles south of the point where it should have been, but it was probably the best Bromwell could do with his equipment in that terribly rough terrain. The point he was supposed to mark was on the east side of the Absaroka Range, across 9,600-foot Eagle Pass.

It is hard to believe that Captain Bromwell did not know the initial point for the east boundary survey was out of place; in any case, that fact seems to have remained undiscovered before the work of surveying the line was begun in 1901 (a hiatus caused by U.S. involvement in the Spanish-American War). Mr. Edward F. Stahle, who completed the surveys on the north, west, and south boundaries that year, estimated that the unsurveyed portion of the east line could be completed for $60 per mile, or $3,000, "since there still remains unsurveyed about 50 miles . . . extending from the monument established by Captain Bromwell, U.S. Engineers, on the boundary east of Yellowstone Lake, to the northern boundary of the park."[23]

The consequences of Captain Bromwell's positioning of the zero-mile marker, which became apparent in 1903, were reported by Captain John Pitcher:

> The entire boundary line of the park has now been surveyed, with the exception of about 6 miles near the northeastern corner. In order to close out this survey it will be necessary to survey about 4 1/2 miles of the eastern boundary and nearly 1 mile of the northern boundary. Fifty miles of the eastern boundary line was surveyed during the past summer by Mr. Edward F. Stahle under a contract awarded him by the honorable Secretary of the Interior. Mr. Stahle commenced his survey at the monument established by Captain Bromwell, and which was reported to have been established at a point 10 miles east of the easternmost point of the Yellowstone Lake; but it was found that this monument was about 4 1/2 miles south of that point, and in consequence the survey of 50 miles of the boundary line northward from this point only reached to a point about east of the head of Amphitheater Creek and nearly 4 1/2 miles south of the north boundary line of the park. . . . The examiner of surveys, Mr. J. Scott Harrison, is at present engaged in finishing this work, having been authorized to do so by the honorable Secretary of the Interior. The completion of this small piece of work will finish the survey of the entire park boundary.[24]

The foregoing summary was necessary to establish three facts: that the stream now shown as Monument Creek was named Mountain Creek in 1885, and was known by that name until 1961; that the first step in the survey of the east boundary of Yellowstone Park was the placing of the zero-mile post by Captain Bromwell in 1894; and that the survey of the old line was completed by J. Scott Harrison in 1904. Now let us consider the name change from Mountain Creek to Monument Creek.

In 1953 Jack E. Haynes, a member of the superintendent's Place Name Committee, wrote a letter to U.S. Cadastral Engineer William

R. Bandy seeking his help in changing the Mountain Creek name to Monument Creek, on the basis of the following information:

> Yesterday I was informed by Lee Coleman, District Ranger, National Park Service, Yellowstone Park, Wyo., who is a friend of John Scott Harrison, who he states was U. S. Cadastral Engineer of this District in about 1931–2, that Mr. Harrison when a young man was engaged in the surveying party that surveyed the East Boundary of Yellowstone National Park. The party of which Mr. Harrison was a member established a monument East of Yellowstone Lake near the mouth of Beaverdam Creek, then proceeded S.E. up the Yellowstone River to the mouth of a creek indicated on the map as 'Mountain Creek' which they ascended until that creek turned northward, from which point they went East up the slope and established a Monument, the *zero* (0) Monument, on the East Boundary of the Park, exactly ten miles east of the eastern extremity of Yellowstone Lake, according to Mr. Harrison's statement, told to Lee Coleman some years ago. Mr. Harrison stated also that the party named the creek *Monument Creek* not Mountain Creek, and that when the USGS map was published that name was changed due, probably to a draftsman's error, from Monument Creek to Mountain Creek.[25]

In view of the published facts concerning the survey of the east boundary, Mr. Harrison's recollection could only be correct if he was a member of Captain Bromwell's party of 1894. At that time the stream name in question appeared on the published topographic quadrangle as Mountain Creek, the name given it by the Geological Survey party in 1885. At the time of Mr. Harrison's conversation with Lee Coleman — about 1950 — he would have been at least 75 years old and likely over 80, which is a long time to remember a minor detail. In the absence of confirming records, and there appear to be none, it is my opinion that Monument Creek should give way to the original name — Mountain Creek.

Other usages involving the name Monument will appear in later chapters.

## LONE STAR

This singularly appropriate name borne by a beautiful geyser may be a relic of that brief period when surveyors of the Northern Pacific Railway were afield in Yellowstone Park — or it may have been given earlier. It depends upon whose story you like best.

As the rails of the Northern Pacific main line approached Benson's Landing on the Yellowstone River in the summer of 1882, a party including two civil engineers, Carl A. Hals and A. Rydstrom, ran an exploratory survey across the Park to determine the practicability of extending a branch line into or through the area. Three miles south of Old Faithful, as they were running their line up the Firehole River, they came upon a remarkable geyser cone in a small opening in the forest. S. P. Panton, who was a member of that party, says they assumed they were its discoverers, "so we named it 'Lone Star Geyser' on our profiles."[26]

The other story is attributed to a Mr. J.M.V. Cochran, an eighty-four-year-old resident of Oak Grove, Oregon. He wrote:

> To the best of my recollection it was in 1874 when I was out with Col. W. D. Pickett of Louisville, Kentucky, on a hunting trip. We were camped near Old Faithful Geyser. I went out hunting for bear and was on a high butte over-looking the basin when I saw the steam from a geyser above the trees and went down to it. There were game trails near but no blazed trails nor trails showing signs of having been traveled by horses, nor any signs of old campfire — in fact, nothing to show that anyone had ever been there before.
>
> The next day I returned with Col. Pickett and took with me a board off a cracker box and nailed it up on a tree and printed on it the name "Lone Star Geyser."
>
> On this trip we did not see anyone in either upper or lower basins but on the way out met Mr. Calfee and Mr. Catlan near Sulphur Mountain. They were on their way into the basins to take pictures of the geysers. I told them about the Lone Star Geyser and gave them directions how to find it from Old Faithful Geyser. If they took views, as they intended, they would be the first ever taken of Lone Star Geyser.[27]

Cochran's presence in the Park with Col. Pickett is confirmed by the latter's book, *Hunting At High Altitudes* — but the year was

1879, not 1874 (hunting was allowed in Yellowstone Park until 1883). Although Henry Bird Calfee did enter Yellowstone Park as early as the summer of 1873, the period from 1877 to 1881 corresponds with his photographic activity there. Calfee's work as known today does not include a photograph of Lone Star Geyser.

Lone Star Geyser. There are two stories explaining how the Solitary Geyser (1872) became the Lone Star Geyser in 1877 or 1882; both are possible.

Returning to events in the year 1882, that September Panton met photographer Frank J. Haynes in the Upper Geyser Basin and showed him the "new" geyser. It was not really new, for the Hayden Survey had found it in 1872 and named it "Solitary Geyser" because of its isolated location. But stereoscopic views of the geyser were sold to thousands of visitors by Haynes, spreading its fame — and it has remained the Lone Star.

The Geological Survey later salvaged the "Solitary Geyser" name. In 1878 the Hayden Survey had noted an isolated thermal feature on the hillside above the Lion Group at Old Faithful and had mapped it as "Solitary Sp." That spring was used as a source of hot water when A. C. Hamilton built his new bathhouse in 1916. As a result of lowering its water level, Solitary Spring became a geyser, and that change was legitimized by renaming it Solitary Geyser — which it no longer is with the bathhouse gone and the water level restored as it was originally. Some things just never seem to work out right!

A further word about that party of railroad surveyors: they left another place name to remember them by. Grants Pass, that easy route between the Madison and Snake River drainages that they found, is named for M. G. Grant, the location engineer of the Northern Pacific Railway — not for President Ulysses S. Grant.

## BECHLER

This surname of a little-known topographer has become so thoroughly identified with features in the southwest corner of the Park as to become its common denominator.

Beyond the fact that he was a topographer with the Snake River division of the Hayden Survey in 1872, Gustavus R. Bechler remains an unknown. Just why Dr. Hayden named such an important feature as a river for him — when he was not even its discoverer — is puzzling. According to Hayden:

> Our hunter, Frank Mounts, had reported a valley about two miles south of us, through which a stream flowed westward. Accordingly, while the train moved about three miles nearly due east, to a new camp on Shoshone Lake Mr. Bechler and I crossed the southern divide to the head of the newly-reported river. We found it to be a large stream, formed, within a short distance, from the abundant flow of numerous large springs bursting from low down in the sides of a high plateau of very porous volcanic sandstones, which we afterward found to extend for several miles southward. . . . In the absence of any prior designation, I have called this Bechler's River.[28]

The name was entered on the 1878 topographic sheet as Bechler's Fork, and was put into its present form by the Geological Survey in 1885.

The southwest corner of the Park went unwatched for a quarter century, during which time poachers and cattlemen used the area as if it were part of the public domain. Bartlett Slough takes its name from a rancher who had his cattle-feeding pens where the slough joins Boundary Creek (he wintered stock there on stacked meadow hay). An early freight road, marked on present maps as "Old Marysville Road," looped deeply into the Park as it passed around the Bechler Meadows. It linked Marysville, Idaho, a former town on the eastern outskirts of Ashton, with Marysvale, Wyoming, the early settlement near present-day Jackson.

Although the need for protection from illegal activities was recognized as early as 1899, only an occasional patrol from the South Entrance or from Riverside Station was possible until 1911, when the Bechler Station was built on Wyoming Creek (mistakenly built there instead of on Rock Creek, at the Park corner). The soldier station became the Bechler River Ranger Station after the army left the Park, and was so shown on maps from 1928 on.

Following World War I there was political pressure to turn the Bechler Meadows into a water storage reservoir for the benefit of agricultural interests in Idaho. Consequent interest in the southwest corner resulted in two more spin-offs of the Bechler name — Bechler Canyon (USGS, 1921), and Bechler Falls, named the same year by William C. Gregg, a friend of the Park who defended it against the irrigation interests.

Strangely, though the Bechler Meadows were known by that name for many years, they did not make an appearance on the Park map until 1957.[29] An administrative use — Bechler River Trail — on the same map completes the usages that have devolved from the Bechler name.

## MISCELLANY

The Hayden Survey and its successor, the United States Geological Survey, have been responsible for approximately 926 Yellowstone place names (an admittedly inexact number compiled from maps of those organizations that undoubtedly includes names from other sources not now identifiable). The names can be grouped as nondescript, characteristic, or determinate, as follows:

The nondescript place names are those for which there is no logical reason; at least, if there ever was a peculiar circumstance behind the naming, it is lost to us now. The name Sunset Lake — given a hot pool in Black Sand Basin by the Hayden Survey in 1878 — made so little sense that the Geological Survey tried to rename it "Sunshine Lake" (1904), but the original remains in use. Similarly, King Geyser in the West Thumb Geyser Basin (1904) could refer to its size (in that area of small geysers), or to a person, with the latter finding some support in a 1960 map that calls it "King's Geyser." The most that can be said for the nondescript names is that they do provide a handy way to refer to particular features, as illustrated by the Specimen Ridge area. There the drainages, clockwise from the west, are named: Agate, Quartz, Crystal, Jasper, Amethyst, Chalcedony, Opal, and Flint (then they had to spoil it with two grasses: Clover and Timothy!). One would be foolish to expect agates on Agate Creek or flint rock on Flint Creek, and there is no more clover on Clover Creek than elsewhere on the Mirror Plateau, but those names do serve to identify particular streams.

The characteristic place names — and this is a large group — are those where some physical attribute of the feature appears to have suggested the name. The qualification "appears" is necessary because the circumstances of the naming are often undocumented. Tangled Creek does have a tangled skein of watercourses, Rainbow Falls does display rainbows in its spray, and there is a notch in the top of Top Notch Peak. Such names are a grade better than the nondescript because they do not raise the haunting question, *Why*?

But it is the small group of determinate place names that provides most of the interesting stories, and some of those are worth considering. The surveyors — like the explorers, prospectors, and trappers — sometimes named features for people. The attachment of the name of James Stevenson, Dr. Hayden's assistant, to an island and a peak, and of Gustavus R. Bechler, a topographer of the Hayden Survey, to a river and by later extension to a number of other features in the southwest corner of the Park, has been mentioned. Other surveyors, however, have been honored on the Park map.

The 1878 field party of the Hayden Survey remembered two of their comrades — George B. Chittenden and William H. Holmes — in that manner. Chittenden was a topographer for the Survey during the five seasons of 1873 through 1877, and Henry Gannett explains the honor accorded him in these words: "Of the prominant [*sic*] peaks of this [the Absaroka] range may be mentioned Mount Chittenden, named for Mr. George B. Chittenden, whose name has long been associated with this survey."[30]

Mount Holmes, that great peak forming a spectacular finial for the southern end of the Gallatin Range, was named by Henry Gannett in 1878 for William Henry Holmes — the artist-geologist for the Hayden Survey (1872–1878). It is obvious from Holmes's report that he was pleased:

> On the morning of the 8th Mr. Gannett and I set out to ascend the southern summit of the range, which is only inferior in elevation to Electric Peak, being 10,100 [10,336] feet in height.

> This peak, or what we suppose to be this peak, has been spoken of occasionally by visitors as Mount Madison, but on what authority or for what reason it is not known. Mr. Gannett having first ascended it and determined its height and geographic position, proposed to give it the name of the writer and by that name it will appear on his maps.[31]

Mount Holmes had previously been vaguely identified as "Mount Gallatin" by Captain Raynolds (1860), and as "Mount Madison" by Norris in 1878, but it was the Hayden Survey's designation that stuck. On the climb to the summit, Holmes found marine fossils on a ridge somewhat below the top; among them were trilobites "occurring in great numbers." That outlier was later named Trilobite Point.

Holmes is also remembered on the Park map by Holmes Point, a slight protuberance in the shoreline of Yellowstone Lake at the southern end of Mary Bay. The initials "W.H.H." found incised on a rocky outcrop there are presumed to be his.

There are other place names of a personal nature worth mentioning. Peale Island, called "Peale's Island" by the Hayden Survey in 1872, was named for Dr. Albert Charles Peale, who served the Survey as a minerologist from 1872 to 1878. He was a grandson of the famous American painter Charles Wilson Peale, and like Dr. Hayden, a qualified medical doctor. The island, which is at the southern extremity of the South Arm of Yellowstone Lake, was the site of a U.S. Bureau of Fisheries station where trout spawn was taken for the hatchery at Lake.

Hague Mountain, which forms the "pommel" of the Saddle Mountain group, was named for geologist Arnold Hague in 1930 in recognition of his many years of service with the Geological Survey. The Glacial Boulder now stranded in the forest on the west rim of the Grand Canyon near Inspiration Point was for a time called the "Hague Boulder," because Arnold Hague identified it as debris plucked from granitic bedrock near the northeast corner of the Park and transported to its present resting place by a river of ice. There was also a very large pine on the Fort Yellowstone lawn (opposite the old Haynes Picture Shop) that local tradition called the "Hague Tree" because he camped there before there was a settlement at that point. The tree died a number of years ago, but its massive stump is still visible.

The name that is first in any alphabetical listing of the Park's place names — Abiathar — was also drawn from a personal name. The Geological Survey (1885) used the middle name of Charles Abiathar

White, one of the Survey's paleontologists (but with no known connection to the Park), as the designation for a rugged peak standing above the Northeast Entrance.

The particular interest of geologists in the thermal features of Yellowstone Park has resulted in many place names of the "characteristic" type. These cover such presumed attributes as: *action* — Atomizer Geyser, Whirligig Geyser, Restless Geyser, Union Geyser (uniting three vents), Surprise Geyser (for its reaction when disturbed), Roaring Mountain (now very quiet), and another deceiver, Constant Geyser (no longer constant); *shape* — Kidney Pool, Oblong Geyser, Round Spring, Medusa Spring, Horseshoe Spring, Flat Cone (from its low profile), and Liberty Cap (from a fancied resemblance to the cap worn by many common people in the French Revolution); *color* — Ebony Geyser, Canary Spring (yellow), Gentian Pool (blue like the flower), Kaleidoscope Geyser (many brilliant colors), and Grand Prismatic Spring, with its rainbow colors reflected in the steam; *ornamentation* — Bead Geyser, Echinus Geyser (with spiny crater), Cauliflower Geyser, Crested Pool, Fungoid Geyser, and Pearl Geyser (with a nacreous lining); *sound* — Black Growler, Whistler Geyser, Locomotive Geyser, and Thud Geyser; *location* — Deep Creek Springs, Half-way Springs, Hotel Terrace, Solitary Geyser, Undercliff Spring, Quagmire Group, and Snake Hot Springs, which are not reptilian — only located on the bank of Snake River.

A complete listing of these "characteristic" place names that have been given to features in the Park's thermal areas — an estimated 150 geysers and 5,000 hot springs, hot pools, and steam vents — would be monotonously long, for the surveyors tried to catalogue all the important ones. However, the circumstances behind the naming are known for only a few.

The best of those place names with a known origin is Chinaman Spring (also called the "Chinaman Geyser" and "Laundryman's Spring"). The name grew out of a happening that made a rather inconspicuous hot spring into a legend known to nearly all Park visitors of yesteryear. It was recorded first by geologist Arnold Hague in these matter-of-fact words: "In the summer of 1885, a Chinaman employed as a laundryman for the accomodation of the tourists at the Upper Geyser Basin accidentally discovered that soap thrown into the spring from which he was accustomed to draw his supply of water produced an eruption."[32]

There is nothing in that statement to indicate that the event was particularly terrifying, or damaging, but it did have two important effects: Hague's scientific curiosity led him into a series of experiments that established the conditions under which soap could induce eruptions in some geysers, and the visiting public's interest in this new development led to a fad called "soaping geysers" — which soon became very unpopular with the authorities. It might be noted that the demand for soap in the geyser basins was all out of proportion to the standard of cleanliness of that day!

It remained for a newspaper columnist by the name of Bob Burdet — a folksy humorist who was widely read — to turn the Chinaman's misadventure into a Yellowstone legend. With journalistic embellishment, his copy came out this way:

> It is written in the Archives of Yellowstone Park that a child of the Flowery Kingdom, wearing the usual smile on his childlike face, and his shirt outside his trousers, came to the Upper Geyser Basin to establish a laundry, because there was enough hot water there to run a Presidential Campaign. He pitched his tent over a thermal spring, wrote his name in weird characters upon a sign board, and when the raiments of the native and the pilgrim came in, he chucked the whole invoice into the bubbling spring. Then he threw in a bar of soap, and smiled to see great forces that upheave mountains and shake continents, and toss the mighty Globe into convulsions most awful, doing plain washing in a Chinese laundry. But this spring was a slumbering geyser. The soap awakened the imprisoned giant; with a roar that made the earth tremble, and a shriek of a steam whistle, a cloud of steam and a column of boiling water shot up into the air a hundred feet, carrying soap, raiment, tent and chinaman along with the rush, and dropping them at various intervals along the way. 'Hookee-la,' said John, when he came down — way down! 'Joss, he no like washee for China-boy. Too muchee-bubblee make.'[33]

Thereafter, the visitor got the story of the Chinaman Spring in many different forms: always with the washing violently regurgitated, usually with the tent blown asunder, and sometimes with the Chinaman going along for the ride. It is even claimed that the Chinaman was so frightened that he disappeared, "never to be seen again in the vicinity." But that is stretching a tale too far, since a lady visiting the Upper Geyser Basin the following year noted that, "The everlasting Celestial has utilized one of the smaller geysers and turned it into a wash-tub." Occupational interest led her to describe the laundry process in greater detail than is warranted here.[34]

It is a tale that has been reshaped in many ways, but one thing has not changed: "soaping geysers" for the thrill of an "on-demand" eruption is still unpopular with the authorities. Even the name Chinaman Spring is out of favor these days, and it seems likely that "Laundrymans Spring," — a less offensive term that still retains something of the historic happening — will replace it. The "Merry Widow Geyser," another feature whose name derives from a somewhat similar happening, will be discussed in Chapter 7 (Miscellany).

Another interesting place name grew out of a discovery by Frank H. Knowlton and Walter H. Weed in the summer of 1888: they found dead bears in a ravine leading down to Cache Creek at Wahb Springs. The place, named Death Gulch by the Geological Survey, is described by Weed as follows:

> The gulch ends, or rather begins, in a scoop or basin about two hundred and fifty feet above Cache Creek, and, just below this, was found the fresh body of a large bear, a silver-tip grizzly, with the remains of a companion in an advanced stage of decomposition above him. Near by were the skeletons of four more bears, with the bones of an elk a yard or two above, while in the bottom of the pocket were the fresh remains of several squirrels, rock hares, and other small animals, besides numerous dead butterflies and insects.[35]

The deaths were attributed to asphyxiation by a toxic gas — probably carbonic acid gas resulting from the action of acid waters on the underlying limestone. In applying the name Death Gulch to this V-shaped rent in the canyon wall, Knowlton and Weed were figuratively linking it to the ancient fable of the "Death Valley" of Java. In that tale there was — in the center of the island — a valley dominated by the deadly *Upas* tree, which poisoned everything within a radius of ten miles. According to Foersch, a Dutch Surgeon, its virulence was so great "there was no fish in the waters, nor any rat, mouse, or other vermin to be seen there, and when birds fly so near this tree that the effluvia reaches them, they fall a sacrifice to the effects of the poison."[36] This account was found by Junghuhn to be a gross exaggeration of a natural phenomenon; the lethal vale was only a saucerlike depression about a hundred feet in diameter and fifteen feet deep — a mere sump collecting carbonic acid gas emanations from a vent. Though the *Upas* tree does have a poisonous sap, it was not responsible for the deadly effect.

This carbonic acid gas is the agent that kills small birds and mammals sheltering in the Stygian Cave at Mammoth Hot Springs, and its presence in the Devils Kitchen led to discontinuance of guided trips that once gave visitors a glimpse into the labyrinth underlying the hot spring terraces. During the summer of 1898, the soldiers stationed at Soda Butte checked the Death Gulch area frequently, reporting the remains of eleven bears on one visit. No carcasses have been reported by the occasional visitors in recent years.

Wahb Springs, where carbonic acid gas bubbles to the surface of Cache Creek below Death Gulch, takes its name from the aged grizzly bear of a fictional account. Ernest Thompson Seton used Death Gulch as the setting for Wahb's melodramatic end in a bearish euthanasia. The naming of the springs by the Geological Survey (1904) followed publication of Seton's *Biography of a Grizzly.*

The name given a feature in the Heart Lake Geyser Basin — Rustic Geyser — is both descriptive and tantalizing, for it exposes a mystery that likely will never be explained. Geologist Holmes described the geyser and its naming in this manner:

> The basin in which the water stands is about 9 feet in diameter, the rim being roughly hexagonal in shape — the result of an artificial enclosure of small pine logs which have been placed thus by visitors perhaps a year since. These logs are entirely covered by a silicious deposit of yellowish color and most beautifully beaded . . . covered to the thickness of an inch or more. The basin rim thus surrounded by this curious sugar coated frame work suggested the name Rustic.[37]

In view of the slow rate at which sinter is known to accumulate in other geyser basins, it does not seem likely that logs showing an inch of deposit could have been in place for only a year. When examined by Allen and Day in 1925, the logs were found to be "encrusted with sinter to a thickness of about half an inch,"[38] which does not agree with Holmes's observation.

Thus, it is quite probable the logs have been in place a very long time — perhaps even as long as the two or three hundred years postulated by T. Scott Bryan in *The Geysers of Yellowstone* (1979). This might even be evidence that prehistoric people were interested in the hot springs.

Congress Pool in the Norris Geyser Basin is a place name that has been traced to Hiram M. Chittenden — this time not as an engineer, but as an author.

The stagecoach road from Mammoth Hot Springs entered the Norris Basin near Nuphar Lake at a point where a steam vent hissed noisily beside the track. In the winter of 1893, the vent closed off and the pent-up energy blew a crater in the road — a violent event that left the road blocked with debris. As the 1894 season opened, the new pool boiled with great violence, promising to be the biggest spouter in the Park. It happened that the Democratic 53rd Congress just then in session also promised great things, and it seemed right to call the new feature "Congress Spring," and — hopefully — "Congress Geyser."

But like the 53rd Congress, "Congress Spring" never made the grade. From a beautiful, pale-blue boiling spring it degenerated into an angry mud-boiler, and finally to a very ordinary pool — sometimes clear, sometimes turbid, and often dry.

That would be enough to say about Congress Pool were it not for a bit of misinformation that requires explanation. The statement made by Allen and Day that "it was named by Hague in honor of the Fifth International Geological Congress which met in Washington in 1902"[39] cannot be taken seriously in view of the many references to Congress Pool during the previous seven years.[40]

In addition to the foregoing place names, both important and unimportant, the surveyors have left us several with no real existence on the ground — though they have proven durable enough on paper. Chief among these nonexistent place names is "Lost Lake," an ephemeral body of water variously placed on Park maps as a feature on or near the Continental Divide between West Thumb and the Shoshone Lake drainage. It is unrelated to the approved place name given to the small lake a half mile above Roosevelt Lodge, or to the unofficial name of the weedy pond that is the remaining vestige of a former fish hatchery near the old powerhouse at Mammoth. This mythical lake — for such it certainly is — had its origin with the Hayden Survey (United States Geological Survey of the Territories) in 1871, and thus was a product of the definitive exploration of the Park.

Dr. Hayden described the lake, which he did not name, this way in his annual report: "We camped at night on the shore of a lake which seemed to have no outlet. It is simply a depression which receives the drainage of the surrounding hills . . . the lake was about two miles long and one wide, and it is doubtful whether it had ever been observed by human beings before."[41]

The Hayden Survey's 1878 Park-wide map shows a smaller (less than a half mile long) pear-shaped lake on the Continental Divide about where the highway crosses, 4.35 miles west of the old West Thumb junction. The lake remained unnamed on that map, but Superintendent P. W. Norris, who undoubtedly took his information from the 1878 map (same positioning and same shape), showed the lake as "Two Ocean Pond" on the map accompanying his 1880 annual report.[42]

Geological Survey maps from the Shoshone quadrangle (1886) through the 1954 revision of the Park-wide topographic map showed the lake as "Lost Lake," but with the same size and shape as on the earlier maps. However, although they kept the lake nearly on the Continental Divide, it was moved a mile south of the highway, where it remained for more than 68 years — until completion of the aerial survey upon which the 1956 West Thumb fifteen-minute quadrangle was based.

The verdict of the aerial camera, which proved that "Lost Lake" never really existed, should have been accepted as final — but it has not been. *Haynes Guide* remained its champion to the last edition (1966), which stated, at Mile 4.20 out of West Thumb: "Earlier maps showed *Lost Lake* about a mile (air-line) southwest of this point on the road. Present maps, made from aerial photographs do not show this lake, which may have been covered with vegetation when photographed from the air." Indeed, a place name study that appeared only last year echoed that thought — "It is possible that Lost Lake, like other glacial lakes, has filled with vegetation."[43] Such suppositions betray a complete lack of knowledge of the process of photogrammetry, which is quite capable of differentiating a vegetation-choked pond from meadow grass, a sagebrush flat or a lodgepole pine forest. These attempts to somehow authenticate the Lost Lake name are based on a reluctance to give up some rather gaudy tales told in past years, particularly those reminiscences gathered by Ranger-Naturalist Trusten Peery in 1937.[44] Those accounts describe a fishable lake visited by different individuals in 1915 and in 1921, but the directions given lead to localities a considerable distance apart in areas subsequently traversed by forest-type mapping crews, blister-rust control parties, and fire fighters — none of whom found the lake — so that it is hard for me to believe it ever existed.

Another of these "lost" features bequeathed us by surveyors — in this instance railroad surveyors — is Arvid Lake. It is shown on the 1882 map of Hals & Rydstrom as being one mile southeast of "Bell's Peak" (present Antler Peak), which would put it on Indian Creek (a stream without lakes). A recent attempt to identify Arvid Lake with a small cirque pond between Echo Peak and the White Peaks is ridiculous; the unnamed pond is not large enough to be a lake mapped as two-thirds of a mile long, and it is too far west by three miles. Arvid Lake, which was to immortalize the first name of surveyor Rydstrom, is simply a nonexistent feature, along with the "Rosa L." and "Twin L." shown on that map. Surveyor Carl J. Hals fared no better than Rydstrom in the matter of a memorial place name; "Hals Lake," identifiable as one of the two where Fawn Creek breaks into Gardners Hole, is now unnamed.[45]

This section can be closed with a misplaced name that is of no particular importance in itself, although important in its relationship to the surveying of Yellowstone National Park. Signal Point — a name now attached to a slight bulge in the shoreline of Yellowstone Lake south of Meadow Creek — originally identified one of the survey stations used by the Hayden Survey in the triangulation for the 1878 topographic map. On that map the station or point from which the angles to other controlling stations were measured was on the little knob 4,500 feet south of the creek mouth and 1,000 feet from the shore of the lake (measured on the present map). The small triangular symbol on the early map indicated the place where a rude tower, fitted with cloth panels visible at a great distance, was erected. Over the years the name designating that survey station was transferred to the insignificant shore feature because that *seemed* right.

## NOTES

1. Ferdinand V. Hayden, *Sixth Annual Report of the United States Geological Survey of the Territories . . . Being a Report of Progress of the Explorations for the Year 1872* (Washington: Government Printing Office, 1873), p. 807.
2. Ibid., pp. 120–121.
3. United States Geological Survey, *Miner, Mont.-Wyo.*, N4500-W11045/15 (1955).
4. Jack E. Haynes, [Reminiscences recorded from 1953 to 1963.] Typed copy in the author's files, pp. 75–76.

5. John Gibbon, "The Wonders of the Yellowstone," *Journal of the American Geographical Society of New York*, 1874, V, p. 136.
6. Hayden, op. cit., p. 55.
7. Philetus W. Norris, *Report Upon the Yellowstone National Park, to the Secretary of the Interior, for the Year 1878*, Dec. 10, 1878 (Washington: Government Printing Office, 1879), p. 995.
8. Jack E. Haynes, *Haynes Guide: Handbook of Yellowstone Park* (St. Paul: Haynes, Inc., 1939), p. 66.
9. "Yellowstone National Park." Quoted in a press release from the office of the Park Superintendent following the death of Lord Dunraven in January 1927.
10. F. V. Hayden, *Twelfth Report of the United States Geological Survey of the Territories . . . for the Year 1878*, 2 Vols. (Washington: Government Printing Office, 1883), II, p. 478.
11. United States Geological Survey of the Territories, *Yellowstone National Park*, topographic map at the scale of 1:126,720, surveyed 1878. Yellowstone Park Reference Library. *Cf.* the crude map tipped in at the back of Norris's annual report for 1878 (cited in Note 7), for an idea of his limited cartographic knowledge in the Mount Washburn area.
12. Philetus W. Norris, *Report Upon the Yellowstone National Park to the Secretary of the Interior . . . 1879* (Washington: Government Printing Office, 1880), p. 29.
13. George L. Henderson, "Wonders in Wyoming," St. Paul, MN, *Pioneer Press*, Dec. 25, 1888.
14. John W. Powell, *Eighth Annual Report of the Geological Survey . . . 1886–87* (Washington: Government Printing Office, 1888), I, p. 152.
15. P. H. Sheridan et al., "Report of Col. O. M. Poe, U.S. Engineers," *Reports of Inspections Made in the Summer of 1877* (Washington: Government Printing Office, 1878), p. 72.
16. William Clark, MS map compiled by William Clark from informants during the period 1806–1811. MS is No. 303-iv, in the William Robertson Coe Collection, Beinecke Library, Yale University, New Haven, CT.
17. A. J. Johnson, *Johnson's Nebraska, Dakota, Idaho and Montana*, [map] Scale: 1 in. = 54 mi., New York, 1865.
18. "A Thrilling Event on the Yellowstone," *The Helena* (MT) *Daily Herald*, May 18, 1870.
19. Hayden, *Sixth Annual Report* (1873), p. 250. See Note 1 for citation.
20. Ibid.
21. *United States Statutes at Large*, Vol. 17, Chap. 24, p. 32.
22. George S. Anderson, *Report of the Acting Superintendent of the Yellowstone National Park to the Secretary of the Interior, 1894* (Washington: Government Printing Office, 1894), pp. 5 & 6.
23. John Pitcher, *Report of the Acting Superintendent of the Yellowstone National Park to the Secretary of the Interior, 1901* (Washington: Government Printing Office, 1902), p. 5.
24. John Pitcher, *Report of the Acting Superintendent of the Yellowstone National Park to the Secretary of the Interior, 1903* (Washington: Government Printing Office, 1903), pp. 4 & 5.

25. Jack E. Haynes, Letter to William R. Bandy, Billings, Montana, June 2, 1953. Copy in the correspondence file, Haynes Collection, Montana State University Library, Bozeman.
26. S. P. Panton, "Early Days in Yellowstone Recalled by N.P. Surveyor," *Wolf Point* (MT*) News*, November 23, 1939.
27. J.M.V. Cochran, "The Naming of the Lone Star Geyser," *Yellowstone Nature Notes*, Oct.-Nov. 1932, p. 47.
28. Hayden, *Sixth Annual Report* (1873), pp. 243–244. See Note 1 for citation.
29. USGS, *Warm River Butte, Idaho-Wyo.*, N4400-W11100/15 (1957).
30. Hayden, *Twelfth Annual Report* (1883), II, p. 483. See Note 10 for citation.
31. Ibid., p. 23. See p. 485 for Gannett's statement.
32. Arnold Hague, "Soaping Geysers," a paper read at the February meeting of the American Institute of Mining Engineers, 1889, and published in *Annual Reports of the Board of Regents of the Smithsonian Institution . . . July, 1892* (Washington: Government Printing Office, 1893), p. 153.
33. Bob Burdet, "In the National Park," *Livingston* (MT) *Enterprise*, July 28, 1888.
34. O.S.T. Drake, "A Lady's Trip to the Yellowstone Park," *Every Girl's Annual* (London: Hatchard's, 1887), p. 348.
35. T. A. Jaager, "A Natural Bear Trap," *Appelton's Popular Science Monthly*, February 1899, p. 475. He is quoting from Weed's article in *Science*, February 15, 1889.
36. F. H. Knowlton, "A Gulch of Death," *Popular Science News* (from an undated clipping in Scrapbook No. 2, Yellowstone Park Reference Library).
37. Wm. H. Holmes, Extracts from the diary of W. H. Holmes, for the winter, spring and most of the summer of 1878, as abstracted by him — probably in 1928. Yellowstone Park Reference Library, p. 10.
38. E. T. Allen and A. L. Day, *Hot Springs of the Yellowstone National Park*, Carnegie Institution Publication No. 466 (Washington: Government Printing Office, 1935), pp. 320–321.
39. Ibid., p. 481.
40. Hiram M. Chittenden, *Yellowstone National Park* (Cincinnati: Robert Clarke Co., 1895), p. 343. See also A. B. Guptill, *Haynes Guide to Yellowstone Park* (St. Paul, MN: F. J. Haynes, 1896), p. 32, and Scrapbook No. 2, Yellowstone Park Reference Library, pp. 49 & 57.
41. Ferdinand V. Hayden, *Preliminary Report of the United States Geological Survey of Montana and Portions of Adjacent Territories, Being a Fifth Annual Report of Progress* (Washington: Government Printing Office, 1872), p. 127.
42. Philetus W. Norris, *Annual Report of the Superintendent of the Yellowstone National Park to the Secretary of the Interior for the Year 1880*, November 30, 1880 (Washington: Government Printing Office, 1881), folded map.
43. Lee H. Whittlesey, *Yellowstone Place Names* (Helena: Montana Historical Society Press, 1988), p. 91.
44. "Lost Lake," *Yellowstone Nature Notes*, XIV/9–10, (Sep.-Oct., 1937), pp. 41–42.
45. See *Map of the Yellowstone National Park, compiled from different official explorations and our personal survey, 1882*, by Carl J. Hals and A. Rydstrom. Scale, one inch = 3 miles. This map was used in several early guidebooks.

# 6

# OFFICIALDOM TAKES A HAND

*Norris*
*Queens Laundry*
*National Park*
*Grant*
*Meldrum*
*Excelsior*
*Virginia*
*Semi Centennial*
*Arthur/Cody/Plentycoups*
*Miscellany*

Almost from the beginning of Park administration, the superintendents and their staffs have influenced the naming of features. This should not be surprising in view of the fact that a superintendent has always been able to support his opinions in published reports, has had a preferential relationship with the survey organizations and the Board of Geographic Names, and is able to influence usages by placing sign boards and through the educational/informational services of his organization. Thus, there is a body of place names that have their origin in official actions.

Place names of this type made their appearance in the superintendency of Philetus W. Norris, who was a pioneer in the fullest sense of the word, and claimed a pioneer's right to bestow place names in the area he labored to open up. He early founded and named the town of Pioneer, Ohio, on land he reclaimed from the Black Swamp, and the former town of Norris, Michigan (now the Hamtramck district of Detroit), was given his name in recognition of his success in making that area habitable through drainage.

In the course of building Yellowstone Park's first roads and trails, Superintendent Norris may have become godfather to more important place names than any other individual. Of one hundred forty-three which can be traced to him, thirty-nine remain in use in unaltered form and at least twenty-one have survived with changes. Survival of so many of Norris's place names was probably due to the signs, or "guide boards" as he called them, which he erected to show the names of features and give directions to travelers. They had much the same effect as cartography has in stabilizing place name usage.

The civilian superintendents who followed Norris were less dedicated than he, but among the assistants — that body of Park police locally styled "rabbit catchers" — were several young men who were interested in place names. Two, Josiah W. Weimer and George L. Henderson, contributed to the legacy of place names, and the latter, while a concessioner, named more features than Norris, though fewer survived (fortunate since his were not as appropriate; this will be considered in the next section).

The acting military superintendents showed little interest in Park place names, being content to leave such matters to the United States Geological Survey during the thirty years of Army administration in Yellowstone.

Management of the area by the new National Park Service after 1916 resulted in renewed interest in place names. An official recognition of the relationship between visitor use and education resulted in establishment of an Education Department (ancestral to all naturalist activities). As has been mentioned in the Introduction, a Place Name Committee consisting of Superintendent Horace M. Albright (chairman), Mr. Jack E. Haynes, photographic concessioner, and Mrs. Marguerite L. Arnold from the Educational Department, took up the task, in 1922, of making order out of years of Topsy-Turvy growth of place names in the Park. Inevitably, they became arbiters of the fate of many existing usages and also originators of a number of new ones.

When Mr. Albright left the Park in 1928 to become the second Director of the National Park Service, that first place name committee became inactive, with Naturalist Newell F. Joyner handling such matters until 1932. Thereafter, place names were the concern of the Park's Chief Naturalist — first C. Max Bauer, then David deL. Condon — until 1957, when the responsibility was shifted to Protection Department personnel.

In retrospect, official place name activity has been less than satisfactory, with some mistakes which might have been avoided had the handling of place name matters followed a consistent pattern.

## NORRIS

There are many usages founded on the surname of the Park's second superintendent, and, though he has been criticized for naming too many features for himself, he did not name as many as his contemporaries thought. Actually, only four — a river, a geyser basin, a mountain and a pass — were given by him; and of those, only two remain in use. That is hardly an excessive egoism in a man who "did the preliminary work with an unselfish devotion and enthusiasm that shortened his life."[1] His predecessor, Nathaniel P. Langford, has characterized him in these words: "He was a good man, a true man, faithful to his friends, of very kind heart, grateful for kindnesses, of more than ordinary personal courage, rather vain of his poetical genius, and fond of perpetuating his name in prominent features of scenery."[2]

Local interests, anxious to end Norris's superintendency for selfish reasons used the newspapers to discredit him by presenting him as an incompetent, doting on his own poetry and, ever busy "bestowing his lordly name upon numerous attractions," which were itemized as "Mount Norris, Norris Geyser Basin, Norris Fork of the Gibbon, Norris Cut-off, Norris Obsidian Glass Road, Norris Plateau, Norris Geyser, Norris Museum, Norris Paint-Pot, Norris Falls, Norris Pass, Norris Pile Spring."[3]

Of course, the foregoing list is ridiculous; a mean composite of mostly non-names and perversions. There were only four real place names among the twelve, and Norris *was* responsible for them, as the following listing shows:

The first, or oldest, of the Norris place names was "Norris Fork," which he gave to the easterly branch of the present Gibbon River in 1878.[4] At that time, neither of the streams converging at the geyser basin had been named, though the Hayden Survey had put Gibbon's name on the lake which is the source of present Solfatara Creek, and Norris may have thought of it as the main course of the Gibbon River. That is not how the United States Geological Survey interpreted the drainage pattern, and, in 1885, the Gibbon River name was continued up the easterly branch to Wolf Lake, doing away with "Norris Fork" as a place name.

The second Norris place name, "Norris Geyser Plateau," made its appearance in 1879.[5] He may not have known that the Hayden Survey had already named that thermal area Gibbon Geyser Basin on its 1878 topographic map (which was only "in press" in 1881 and not yet available). However that may be, Norris changed the form of his usage to Norris Geyser Basin in 1881,[6] and that form, confirmed by the United States Geological Survey in 1885, has remained in unquestioned use.

The third Norris place name, "Mount Norris," also made its appearance in 1879,[7] but the reference is so vague it cannot be identified with certainty; however, this first "Mount Norris" was probably the peak Norris called "Bell's Pk." the following year (in honor of an assistant secretary of the Interior Department), and which is now known as Antler Peak. The best guess is that Norris changed the name when he found that the Hayden Survey had already put his name on a summit of the ridge between Soda Butte and Cache creeks in the northeast corner of the Park.[8]

The fourth, and last, of the Norris place names, Norris Pass, appears to have been a by-product of the success of Luther S. "Yellowstone" Kelly in finding a practicable route across the Continental Divide. Norris had asked Kelly, who was really an Army scout, to guide the escort and pack train of R. J. Reeves, the surveyor then running the western boundary of Wyoming Territory, across the wilderness between the upper Yellowstone River and the geyser basins on the Firehole. Of this Kelly wrote: "Mr. Norris, the superintendent, immediately dispatched a construction party to cut out and build a temporary roadway along the course of our trail. I say he hastened to do this, fearing, mayhap, that otherwise the trail might disappear and leave no

Mount Norris (9,936 ft.). Named for the second superintendent of Yellowstone National Park, Philetus Walter Norris, who opened it for the use of the American public.

trace to guide the eager tourists whom he expected in shoals to view the sights of this wondrous region."[9]

Unfortunately, Norris's account of the discovery of the trail, and the pass, which he claims to have made after several futile searches for a practicable route, does not agree with Kelly's version. Here is how Norris first recorded the finding of this important link in the Park's developing system of trails:

> Thus baffled as had been all previous explorers, I, as a last effort, ascended the north bank of the Fire Hole River, past a succession of enchanting rapids, cascades, and cataracts, and after some days of excessive exposure and hardship, and nights of sleepless cold and anxiety, was rewarded by more than hoped-for good fortune in tracing a chain of small mountain and timber-hidden parks, and lovely lakelets, to a narrow timber-hidden, but deep, direct, and excellent pass, which within less than 2 miles crosses, or rather cuts, the main divide from the Fire Hole waters of the Atlantic to a sheltered grassy park near Shoshone Lake, and the spring and rivulet fountain heads of the Columbia of the Pacific.[10]

Another version appeared in 1884 with the publication of Norris's book, *The Calumet of the Coteau.* Writing of the pass, he says: "This is the low, direct, and short but muddy Norris Pass, thus named by the famous scout and guide, Yellowstone Kelly, upon its discovery by myself, after fruitless search by himself and others for any pass in that vicinity."[11]

In most instances where Norris named a feature, he was quite forthright about it; but in this case, the pass went unnamed in his first description, and it is named only in the table of trail distances which accompanied the 1879 *Report* (see p. 29) and subsequent reports. The name does appear on several maps Norris prepared later, and in the book published following his superintendency. Whether it was Kelly or Norris who discovered this important pass probably is beyond proof, yet I tend to think that it was not Norris; that he claimed a discovery he did not make because he had put so much effort into the search and wanted so badly to succeed. Regardless, the name Norris Pass has remained in use through the years.

The foregoing analysis of the four Yellowstone place names Norris gave in honor of himself can be summarized this way: two of them, "Norris Fork of Gibbon River" and his "Mount Norris," did not survive (the present Mount Norris was named by the Hayden Survey *for* him, not by him), and two of them did survive — Norris Geyser Basin (which was first "Norris Geyser Plateau"), and Norris Pass.

It was unfair of Historian Chittenden to accuse Norris of naming yet another summit for himself, present Dunraven Peak (a statement which is entirely unsubstantiated); nor was it right to charge that "an uncertain number of other features of the Park, were thought by Colonel Norris deserving of this distinction."[12] Norris did name many other Park features, but not for himself.

Before taking up some of the other place names Norris gave to Park features, a few words are in order concerning names he is *not* responsible for. As previously mentioned, a number of real-sounding place names which appeared in the Bozeman, Montana, newspaper in 1880 were intended only as an embarrassment to Norris; they were "Norris Cut-off" (that road wasn't built until 1886, and it didn't become a "cut-off" until the Grand Loop was completed in 1903), "Norris Falls," "Norris Geyser," "Norris Museum" (there was none until 1928), "Norris Paint Pot," and "Norris Pile Spring." The "Norris Obsidian Glass Road" was merely a perversion of the name — Norris Road — which the Hayden Survey applied, in 1878, to the road Norris built that year from Mammoth Hot Springs to the Upper Geyser Basin, and "Norris Park" was merely a caustic intimation that Norris had taken the national park for his own. Over the years, other usages involving Norris's name evolved: "Norris Slide," an uncomplimentary reference to a section of road near the Riverside Mail Station (1881); "Norris Range," G. L. Henderson's attempt to put a name on the height called "Mount Schurz" by Norris (1887); "Norris Hotel," first used for the structure built in 1883, but also applied to later buildings; "Norris Lunch Station," a designation for several temporary structures; "Norris Station," the military outpost (1886–1916), and "Norris Ranger Station" (1917–1959). Norris Junction has been in use as a place name from 1886 to the present.

## EXCELSIOR

Of all the place names Superintendent Norris gave to Park features, this one is most apt; a quite fitting designation for a mighty exhibition of thermal power.

As early as 1869 there was a hint that a very powerful geyser existed in what is now the Midway Geyser Basin. David E. Folsom, describing that area as he saw it October 2, 1869, says, "the river was blood warm a quarter of a mile below. One of the springs was two hundred and fifty feet in diameter, and had every indication of spouting powerfully at times."[13]

Walter Trumbull of the 1870 Washburn Expedition made a similar observation, but, neither the Washburn explorers nor the first Hayden Survey party (1871) observed eruptive action at the huge "caldron" that was pouring so much hot water into the Firehole River. In 1878, the Hayden Survey upgraded Dr. Hayden's earlier descriptive term to a place name — "The Caldron"[14]; but without having witnessed an eruption.

But Superintendent Norris soon had evidence of powerful geyser action:

> Although noted for the deep ultramarine blue, ever-agitated waters, so characteristic of the true geyser when not in eruption, there was neither evidence nor indications of recent eruptions until late in August, 1878. I then distinctly heard its spoutings when near Old Faithful, 6 miles distant, but arrived too late to witness them, though not its effects upon the Fire Hole River, which was so swollen as to float out some of our bridges over rivulet branches below it. Crossing the river above the geyser and hitching my horse, with bewildering astonishment I beheld the outlet at least tripled in size, and a furious torrent of hot water escaping from the pool, which was shrouded in steam, greatly hiding its spasmodic foamings. The pool was considerably enlarged, its immediate borders swept entirely clear of all movable fragments of rock, enough of which had been hurled or forced back to form a ridge from knee to breast high at a distance of from 20 to 50 feet from the ragged edge of the yawning chasm.[15]

Norris also failed to observe eruptions which evidently occurred in September and November of 1880, but, George W. Marshall, hotel-keeper at the primitive hostel in the Lower Geyser Basin, definitely identified regular eruptions beginning in February, 1881. During much of the summer the eruptions were "simply incredible, elevating to heights of from 100 to 300 feet sufficient water to render the rapid Fire Hole River, nearly 100 yards wide, a foaming torrent of steaming hot water, and hurling rocks of from 1 to 100 pounds weight."[16]

Superintendent Norris noted that, by September, 1881, the geyser, "now seems to be settling down to regular business as a two or three hour geyser, so immeasurably excelling any other, ancient or modern, known to history, that I find but one name fitting, and herein christen it the'Excelsior' until scientists, if able, shall invent a more appropriate one."[17]

The Bozeman, Montana, newspaper, never a friend of Superintendent Norris, championed a different name for the great geyser, calling it "The Sheridan Geyser." That is not surprising since General Philip H. Sheridan, the hero of the Civil War battle of Winchester, was very popular; what is puzzling is that both the discovery and the naming were attributed to Jack Baronett, who was on the other side in the Civil War! "Yellowstone Jack" was an officer in the First Texas Cavalry during that struggle, and yet, when General Sheridan visited the Park on his grand junket in 1882, it was Baronett who was his preferred guide.[18]

General Sheridan's name did not "take" where the geyser was concerned, which was probably just as well since he had been honored adequately by the attaching of his name to the outstanding peak which overshadows Heart Lake. For a time, the trail by which Sheridan's party entered the Park from Jackson Hole was also known as "the Sheridan trail," but its importance was lost when a road was built to the south gate.

The last year in which the Excelsior Geyser is known to have erupted is 1890 (noted by Edwin Linton in *Science*, December 28, 1928). However, it remained on Park maps as a geyser until 1927, when the Place Name Committeee proposed a change to Excelsior Geyser Crater, which is the way it is shown on current maps. But in the days of its greatness, when it was very great indeed, Excelsior gave tourists something to write home about, as Margaret Manton did in 1891: "This geyser only plays every other year, and then once every twelve hours it throws a tremendous volume of boiling water 250 feet. Near by is 'Fire Hole' river, which, with each eruption of this geyser, is raised two feet, and the water is made so hot that a soldier who tried to ford the river scalded his horse to death."[19]

Before leaving this vicinity, a few words about the naming of Midway Geyser Basin. The first three names given this thermal area were all products of the Hayden Survey: "Half-way Springs" (1872), because of their position between the lower and upper geyser basins on Firehole River; "Great Springs" (1878), for the size of Cliff Caldron

[Excelsior] and Grand Prismatic Spring; and "Egeria Springs" (1878), the latter a Classical allusion to the mythical advisor and companion to the second king of Rome. It was under this latter name that the Hayden Survey mapped the area.[20]

Others with less appreciation for Classical literature, and greater awe for the terrible forces displayed there, preferred a nomenclature more to the point. M. J. Lott used "Devil's Half Acre"(1880), and Norris used a variation on that theme — "Hell's Half Acre" — in 1882. The United States Geological Survey solved the problem with finality in 1885; the area has been known as Midway Geyser Basin (a Norris name from 1881) ever since.

## QUEENS LAUNDRY

This poetic name, attached to a large hot spring and the nearby ruin of a log bathhouse, is more than a mere place name; it is also a unique testimonial to the vision of Superintendent Norris.

The log bathhouse which stands roofless on the edge of Sentinel Meadow, far from any other structure, is now the oldest building in Yellowstone Park and also the first building erected by the government, solely for the benefit of visitors, in any park. This "Queen's Laundry Bathhouse" was the practical outcome of an event

Queens Laundry Spring. The two-room log bathhouse standing near the spring is the first public-use building built by the federal government in any national park.

which occurred during construction of a "road" (better described as a trail) from the Riverside Mail Station on Madison River to Marshall's mail station near the mouth of Nez Perce Creek in 1880.

> It was while thus engaged that, during a Sabbath's rest and bathing recreation, some of the boys crossed from our camp to the attractive bordered pools below this great boiling fountain, and in one cool enough for bathing discovered its matchless cleansing properties, and from the long lines of bright-colored clothing soon seen drying upon the adjacent stumps and branches, while their owners were gamboling like dolphins in the pools, the envious cooks and other camp attaches dubbed it the Laundry, with a variety of prefixes, of which that which I deemed most appropriate adheres, and hence the name Queen's Laundry.[21]

The next season — summer of 1881 — Norris returned to the site and began construction of a "double-roomed, earth-roofed bath house. . . together with wooden troughs for conveying water thereto for the use of the public."[22] My examination of the old bathhouse in 1961 showed the building to be in remarkably good condition for its 79 years; however, it was also obvious it had never been completed, and probably saw little use.[23]

There is no evidence that the building had a name, other than "bath house" until 1961; in fact, it appears to have been forgotten after Norris was replaced, as superintendent, by Patrick H. Conger on March 1, 1882. It is that neglect and its isolation which has preserved it as a relic from the Park's earliest days.

## VIRGINIA

This given name was probably applied to a Park feature in an attempt to curry favor, but it has since been incorporated into a number of place names.

A road was constructed in 1886 to connect Norris Geyser Basin with the Grand Canyon area, which previously could only be reached from the Lower Geyser Basin by way of the difficult road over Mary Mountain. The new route, later called the "Norris Cutoff," went up Gibbon River, passing a beautiful cascade which had no name up to that time. E. J. Lamartine, the foreman of a crew employed in building the road, has been credited with naming the cascade for the daughter of Sir Charles Gibson, at that time head of the Yellowstone Park Association.[24]

It should be noted that early editions of Chittenden's book describe Virginia as Gibson's wife, an error corrected in the fourth, where she appears as his daughter. No reason is given for the naming, but it is tempting to ascribe it to Lamartine's practice of currying favor with all superiors (a habit which, coupled with poor job performance, eventually got him fired).

Why a middle-aged road foreman should name a beautiful cascade for a young lady so patently out of his class is an interesting puzzle, but it is less important than the fact that the name has proliferated. The canyon through which Gibbon River flows below Virginia Cascade was called "Virginia Canyon" in the guidebook published by F. J. Haynes in 1896; however, the reconstruction which took the road out of the canyon bottom in 1902 also de-emphasized the canyon and it was dropped from the guidebook as a place name.

The extensive meadow above Virginia Cascade was referred to in the information booklet for 1923 as Virginia Meadows,[25] and that usage was confirmed on the topographic quadrangle,[26] *Norris Junction, Wyo.*, 1958. The latest spinoff from the Virginia name has been the unofficial use of "Virginia Lake" to designate a small body of water about two and a half miles southeast of the meadow.

## NATIONAL PARK

This term, which is very American in its origin, has been used in Park place names both in a serious way and lightly.

Something must be said of the origin of this term as a background for its use in Yellowstone Park. That it was an expression of an idea taking shape quite early in American thought is evident from several strikingly similar usages. The artist, George Catlin, travelled up the Missouri River to Fort Benton in 1832 on the steamboat *Yellowstone*, and it was that experience which led him to suggest that the western plains, from Mexico to Canada, should be reserved as "a nation's park, containing man and beast, in all the wild and freshness of their nature's beauty."[27]

In speaking of the grant of the Yosemite Valley and Mariposa Big Tree Grove to the State of California for park purposes in 1864, Prof. Josiah Whitney claimed the area had been made "a National public park . . . under the charge of the State of California."[28] Thus, the Yosemite Grant, though not a Federal area, was beginning to be thought of as if it were.

It remained for a Washington, D.C., newspaper to make the first use of the term exactly as it would later appear in Yellowstone

National Park Mountain (7,560 ft.). As seen from the meadow where the Washburn party camped the evening of September 19, 1870.

usage. In an article titled "A National Park For Washington," the *Daily Morning Chronicle* of January 20, 1871, noted the city's "need of a grand national park"; an idea which eventuated as the "National Zoological Park" in 1889.

The term "National Park" was used in referring to the Yellowstone area almost as soon as the idea of reservation became a legislative proposal. In its issue of January 1, 1872, the Virginia City, Nevada, *Territorial Enterprise* noted: "The Hon. N. P. Langford of Montana, the leader of the famous Yellowstone Expedition of 1870, and several scientific and literary gentlemen, is engaged in an effort to have the Yellowstone region declared a National Park."

Only days before reservation was accomplished a Montana newspaper used the term "National Park" in speaking of the passage of the Senate bill by the House of Representatives,[29] and *The Nation*, in its issue of March 7, 1872, also used the term "national park" in commenting on the Park's organic act and its significance.

Thus, the new area had a generic denomination before it had a name. As noted in Chapter 1, the Park was created nameless and did not become Yellowstone National Park until May 10, 1872, when Acting Secretary of the Interior B. R. Cowen wrote Nathaniel P. Langford informing him he had been selected for the superintendency (*Annual Report* for 1872).

That is the background of a term which has formed part of a number of place names over the years, though only two have persisted. First there was "National Park Avenue," an 1891 characterization of the then *new* road across the Lower Geyser Basin — the one later called "Fountain Freight Road" and now a dead-end scenic drive. That road, which was better aligned than most then in use, brought out the grandiose in George L. Henderson's knack for names, but, like so many of his, it didn't catch on.

"National Park Game Enclosure" was used by F. J. Haynes in his 1896 guidebook as a name for the large corral built in the fall of 1895 by Uncle Billy Hofer with the $3,000 of Smithsonian funds furnished by Professor Langley, who hoped to "save" the American bison by enticing a small herd into an enclosure where the animals could be fed in winter and be safe from poachers. Several square miles of the high ground between Trout and Alum creeks in Hayden Valley were stoutly enclosed, but no buffalo could be induced to take up a residence there. After several years of fruitless effort, the project was abandoned and the name faded, though remnants of the fence remain.

National Park Mountain is a name which has persisted, even though the factual basis of the usage is questionable. It originated with Nathaniel P. Langford, a member of the 1870 Washburn party, who claimed in his 1905 account of that exploration that the idea to establish Yellowstone Park had its genesis nearby, in a campfire discussion on the evening of September 30, 1870. Langford has this to say about the event and the naming: "By direction of Major Hiram M. Chittenden there has been erected at the junction of the Firehole and Gibbon rivers a large slab upon which is inscribed the following legend:

JUNCTION
of the
Gibbon and Firehole Rivers,
Forming the Madison Fork of the Missouri.

---

On the point of land between the tributary streams, September 19, 1870, the celebrated Washburn Expedition, which first made known to the world the wonders of the Yellowstone, was encamped, and here was first suggested the idea of setting apart this region as a National Park. On the south bank of the Madison, just below the junction of these two streams, and overlooking this memorable camping ground, is a lofty escarpment to which has appropriately been given the name 'National Park Mountain.'[30]

Unfortunately, Chittenden did not have all the facts when he had the sign placed at Madison Junction; only those furnished him by Langford, then busy writing his "diary," and Langford was mistaken in several respects: The Washburn party was neither "first" to make the Yellowstone wonders known to the world, nor "first" to suggest reservation in the public interest. It was not even *the* effective agent in setting in motion the sequence of events which matured in reservation (or national park status).

The fur trappers and the prospectors had accumulated a considerable body of information concerning the Yellowstone region and much of that had appeared in print. Even if that is discounted because of its fragmentary nature, then it still remains that the Washburn party was only one of three expeditions, operating in three successive years, which accomplished the "definitive exploration" of the area and thus truly made the Yellowstone wonders known to the world.

The first of those parties — Folsom, 1869 — brought back information which allowed them to correct the current map so that it provided a reasonable impression of the area's geography, to publish a descriptive account in a national magazine, and, most important, it excited interest, locally, in further exploration. The second party, the Washburn in 1870, was larger and had access to newspapers and magazines which greatly increased dissemination of their accounts. But it was the combined Hayden/Barlow effort of 1871, with its scientific character and official status, which entirely convinced the American public that the Yellowstone region was, indeed, a wondrous place.

Considering the claim that Hedges's campfire suggestion at Madison Junction was the first expression of concern for preservation of the Yellowstone wonders for the benefit of the public, that ignores the fact that the acting governor of Montana Territory, Thomas F. Meagher, made just such a suggestion in 1865, and did so in the *presence* of Cornelius Hedges, and that David E. Folsom made a similar suggestion to Henry D. Washburn following the return of the Folsom party in 1869.[31] Thus, the claim that a campfire at Madison Junction was the birthplace of the national park idea is unsupportable, and the suggestion which is purported to have originated there (no support for Langford's claim is found in the extant diaries of members of the party) could be, at most, only a restatement of those earlier proposals by Meagher and Folsom.

The "effective" suggestion, and one quite distinct from the "first," was made by Judge William D. Kelly, acting primarily in the interest of the Northern Pacific Railroad Company.[32] From the foregoing, it appears that the National Park Mountain name is poorly founded. It would have been better to have left those crags as the "Lookout Cliffs," which is what Superintendent Norris called them in 1880, or even as the "Outer Guard," — the choice of C. C. Clawson in 1871 (he recognized that ridge as part of a gateway to the Firehole geyser basins).

One other usage — National Park Range — promoted by historian H. M. Chittenden in 1917, was merely jousting with windmills, (he had a strong dislike for the phonetically difficult term, Absaroka Range, which the United States Geological Survey had settled on, as noted in Chapter 1; but it was already too firmly established for him to change the usage).

## SEMI CENTENNIAL

This name, used by Superintendent Horace M. Albright for a geyser which he promoted as appearing in the fiftieth year of Yellowstone National Park's formal existence, has been disappointing; a "one night stand" in show parlance.

The Superintendent's *Annual Report* for 1922, p. 27, has this to say about the appearance and naming of the geyser: "On August 14th, one of the Roadside Springs had a violent eruption, threw out quantities of rock and mud, and deluged the road. More activity has since been shown, and has created a great deal of interest. It has been suggested that, if this geyser action continues, it be called 'The Semi-Centennial.' This is the geyser that attracted so much interest as to be discussed in newspapers in all sections of the country."

While engaged in research in the historian's office of the Church of Latter Day Saints, Salt Lake City, Utah, I found a surprising confirmation of the widespread interest in the birth of this geyser. In the Church Journal History, under that same date — August 14, 1922 — was the entry, "New geyser spouts in Yellowstone Park." And yet, this geyser was not all that "new" in 1922. An entry in the Superintendent's Monthly Report for September, 1918, noted that a "large mud spring, 15 miles south of Mammoth Hot Springs, close to the main road, was seen to play thin mud to a height of about 50 feet for five minutes on September 14th." Park maintenance employees had observed the geyser in action on four other occasions that summer.

The eruption which gave Semi Centennial Geyser its name was a furious explosion, estimated at 300 feet high, and it killed trees in an area several hundred feet across, even splashing some cars with its muddy discharge. During the remainder of the season smaller eruptions occurred at lengthening intervals, ceasing entirely before the end of the year. Since that time the dormant geyser has been only a dark pool by the roadside a half mile north of Roaring Mountain; marked for many years by a green-and-white enamel sign describing its transient glory.

Ed Moorman, a Yellowstone old-timer, has claimed that Jack Haynes named Semi Centennial Geyser, but he is wrong in that; Jack set the record straight in the 1939 edition of his guide book by stating it "was named in 1922 by Horace M. Albright then park superintendent."[33]

## GRANT

This usage, which has supplied a name for a new settlement on the south shore of West Thumb Bay, honors the eighteenth President of the United States, Ulysses S. Grant, who signed the Act of Congress creating Yellowstone National Park on March 1, 1872.

The idea of substituting an entirely new settlement for the old development cluttering the thermal area at West Thumb Junction was an idea that grew slowly during the post World War II years. A topographical survey of the area south of Thumb Creek, between the old highway and the shore of Yellowstone Lake, was begun in the fall of 1947. After another season of work the project was allowed to lapse until 1956, when it was resumed under the Mission–66 program of Park development intended to put Yellowstone's facilities in satisfactory condition by 1966.

The new area, which gradually expanded to include most of the land immediately adjacent to the south shore of West Thumb Bay, opened for public use — as a campground — in the 1962 season, though shown on the 1961 edition of the park-wide topographic map as Grant Campground. The name, Grant Village, was probably adopted in 1960, but that point is not clear. It appears in print for the first time in the 1961 *Haynes Guide*, in these words: "Site of Grant Village on shore of the lake, is in process of development for 1962 [and] will include a large campground, trailer village, launching ramps, visitor center and all tourist facilities — a model development."[34]

The foregoing statement was overly optimistic, for the transfer of facilities and services from the old area to the new went forward grudgingly over many years and was not quite complete in 1987.

Grant Village. The visitor center at the new area on West Thumb Bay, named for President Ulysses S. Grant, who signed the act of Congress creating Yellowstone National Park on March 1, 1872.

President Grant has been honored with another place name in Yellowstone Park. In 1931 the United States Cadastral Engineer, William R. Bandy, then engaged in running the new eastern boundary of the Park put the President's name on Grant Peak, a summit overlooking the Little Lamar River. As mentioned in the previous chapter, the Grants Pass name has no connection with Ulysses S. Grant (it was given for M. G. Grant, a Northern Pacific Railroad location engineer).

## ARTHUR/PLENTYCOUPS/CODY

The only thing these three place names have in common is the motive behind the naming; a better relationship with neighbors east of the Park.

The Absaroka Range is a mountain barrier which formerly blocked access to Yellowstone Park from the east for as much as seven months of the year, and that led to a feeling in Wyoming's Park County, and particularly in the gateway town of Cody, of having been cheated, even if only by geography, of something of its natural heritage. Superintendent Roger Toll was aware of that attitude and was anxious to improve the Park's relationship with its neighbors.

Thus, when the survey of the new eastern boundary showed several peaks in need of names, he had an opportunity to salve local feelings. A dedication ceremony was held on the East Entrance road July 3, 1932, under the auspices of the Cody Club, to recognize the naming of three peaks: Arthur, for the first President to visit Yellowstone Park (1883); Cody, for William F. Cody, better known as "Buffalo Bill," and Plentycoups, for a noted chief of the Crow Indian tribe.[35]

The name given the peak which overlooks the East Entrance was particularly pleasing to the people of Cody, Wyoming, since it honored the man for whom their town was named, but it ignored the fact that the peak already *had* a name. The cadastral engineer who surveyed the new east boundary, 1929–1932, named that summit "Mather Peak" in 1931, for Stephen T. Mather, first Director of the National Park Service (1917–1929).[36] Steve Mather's name has never been restored to the Park map, though some less worthy persons have since been immortalized upon the heights.

## MELDRUM

With this surname, the Park has honored both a long-time representative of legal processes and the coming of effective law and order.

Judge John W. Meldrum, first United States Commissioner under the provisions of the Lacey Act of May 7, 1894, was the Park's magistrate for more than forty years, resigning his position July 2, 1935, because of infirmities of old age.

Judge Meldrum was born in Caledonia, New York, September 17, 1843. He learned the wagonmakers trade as a young man and joined the Union Army in 1863, serving with the 14th New York Heavy Artillery until he was discharged in September 1864. Following a long convalescence, he found work as a carpenter at the Little Rock Arsenal, but, upon marrying in 1867, he moved to Wyoming Territory and ranched until 1870. A short period running a meat market followed before he entered politics as Clerk of the Circuit Court, and that turned out so well he was acting governor of the Territory at the time statehood was obtained. Having "read" law, Meldrum was appointed United States Commissioner for Yellowstone Park by Judge Riner upon establishment of that position on a permanent basis with passage of the Lacey Act in 1894. Judge Meldrum held office during a critical period in the Park's history and was the only United States Commissioner authorized to use a rubber stamp in lieu of a signature,

Judge John Meldrum. As the first United States Commissioner under the Lacey Act of 1894, he served forty years as Yellowstone's committing magistrate. He is honored by Meldrum Mountain in the Gallatin corner of the Park.

as his crippled arm prevented him from writing. He died February 27, 1936.[37]

Judge Meldrum's position as first Commissioner under the Lacey Act "To protect the birds and animals in Yellowstone National Park, and to punish crimes in said park, and for other purposes,"[38] has frequently been misunderstood. As early as 1879 an attempt was made, by Superintendent Norris, to have a commissioner appointed, but that, like the similar effort made by his successor, went awry. An appointment was finally secured in the spring of 1891, by which Acting Assistant Surgeon Samuel S. Turner, a contract doctor serving the Fort Yellowstone garrison, was able to act as a committing magistrate after June 26. He was followed by Ellery C. Culver, a park transportation company employee, who served in the same capacity after April 12, 1892. But both of these early commissioners were hampered by lack of statutory definition of misdemeanors and the allowable punishments, and it was that defect which was remedied by the Lacey Act.

With the appointment of Judge Meldrum as a full time magistrate, Law and Order became fully effective in Yellowstone National Park, and it is that fact which was celebrated in 1962 by giving a turret-shaped peak between the forks of Specimen Creek the name of Meldrum Mountain.

## MISCELLANY

It has already been stated that the Park's second superintendent, Philetus W. Norris, was responsible for many of the place names now in use — at least 60 by my count — and he probably named a greater number of the Park's important features appearing on current maps than any other one person. A few of the names given by Norris have been considered earlier; but there are others requiring at least a brief mention.

The influence of Superintendent Norris on the naming of the Park's important features is best seen by listing those not yet discussed in four groups: descriptive, thermal, personal and historical.

Norris's descriptive place names make up the largest group. Of approximately twenty-two names not yet discussed in this section, fifteen remain in use in unaltered form. They are Antelope Creek, Beaver Lake, Blacktail Pond, Bridge Bay, Grizzly Lake, Lookout Point, Obsidian Cliff, Pleasant Valley, Red Rock, Rustic Falls, Saddle Mountain, Swan Lake and Willow Park. Another seven are used in an altered form: "Cañon Creek" has become Canyon Creek; "Devil's Cut" is just The Cut; the ponderous "Grand Canyon of the Gibbon" is merely Gibbon Canyon; "Hoodoo Mountain" is a less pretentious Hoodoo Peak; the poetic "Pass Lovely" has been inverted, as Lovely Pass; "Pelican's Hill" is Pelican Cone, and the "Sheepeater Cliffs" have been generalized as Sheepeater Canyon.

In addition to several names of thermal features already mentioned, Norris originated at least ten others which remain in use. The unaltered names in this group are Big Cub Geyser, Constant Geyser, The Cubs, Devils Thumb, Splendid Geyser and Vixen Geyser; while another four are nearly the same in their present form. "Emerald Geyser" is Emerald Spring at Norris Geyser Basin; "Monarch Geyser" is now only Monarch Geyser Crater; "Monument Basin" is Monument Geyser Basin, though its feeble thermal activity hardly warrants the expanded label, and the "Porcelain Vale" is now Porcelain Basin.

Norris gave personal names to many features. It was evidence of his essential generosity that he remembered friends in that manner, not even forgetting the workmen who labored on his projects. Of the twenty-seven names in this group (not counting the few features he named for himself), nine have survived. Calfee Creek, Miller Creek and Parker Peak honored his companions on the 1880 exploration of the Hoodoo Basin in the Absaroka Range. They were Henry Bird Calfee, a Bozeman photographer; Adam "Horn" Miller, an old prospector with whom Norris had explored on several occasions, and William Heman Parker, a trusted employee.

Information concerning the unusual nature of the "Goblin Land" had come to Norris through prospector Miller, who had been chased out of that area by hostile Indians in 1870. Miller had attempted to take Norris there in 1878, but fear of Indians caused them to turn back. So, the superintendent promptly took advantage of another opportunity to visit the region in 1880. "Finding at the forks of the Yellowstone the noted guides Rowland and Miller, with the famous photographer of the Park, Mr. H. B. Calfee, my assistant, Mr. W. H. Parker, and myself joined them, and with them constituted the first party of visitors to the famous and exceedingly interesting medicinal springs, Hoodoo Basin, the matchless dikes and other unique marvels of the East Fork regions."[39]

Two other personal names appearing on Park features are reminders of a visit by the governor of Wyoming Territory in the summer of 1881. The purpose of the exploration was to discover a practicable route for a wagon road connecting the territory with the Park either by Two Ocean Pass or the new pass immediately east of Yellowstone Lake (which became Sylvan Pass in 1885). The party of Governor John W. Hoyt included Colonel Mason of his staff and his 12-year old son, Kepler, and a note in the Bozeman newspaper states that there were also "fifteen California railroad magnates" accompanying the governor, who intended to "view out a railway from the west."[40]

Superintendent Norris commented favorably upon the resulting report of Governor Hoyt and others, and that did not endear him to the people of Bozeman, who wanted a connection to the Park to be built by way of the Gallatin Canyon from the Northern Pacific Railroad then advancing up the Yellowstone Valley (this had more than a little to do with Norris's eventual loss of his position as superintendent of the Park).

Returning to the place names which are the only lasting legacy of Governor Hoyt's visit, one was Hoyt Peak, which flanks Sylvan Pass on the north, and the other was a wild cascade on Firehole River a short distance above Old Faithful, which Norris felt justified in calling Kepler's Cascade, for "the intrepid twelve-year-old son of Governor Hoyt, of Wyoming, who shared all the hardships, privations, and dangers of the explorations of his father."[41] This place name is officially used as Kepler Cascades.

Superintendent Norris honored his loyal assistant, Clarence M. Stephens, by putting his name on two features within the Park and one which was then outside. Only one remains in use on current maps, Stephens Creek, a stream in the Game Ranch Addition to the Park (two and one-half miles west of Gardiner), where Stephens purchased the ranch of James Henderson. The Stephens place names which did not survive were "Mount Stephens" (1879), now Prospect Peak, and "Stephens Range" (1880), now the Washburn Range.

Stephens was a young civil engineer who came to the Park with Norris in 1879, remaining to the end of his administration and a little more (he stayed to turn over the government property to Patrick H. Conger, who succeeded Norris in 1882). Stephens served as overseer of road construction parties and also wintered at Mammoth Hot Springs as the superintendent's representative in the Park during Norris's absence in the East. He was the first postmaster there.

Something needs to be said concerning those other place names of a personal nature which did *not* persist. "Bell's Lake," a pond north of Mount Washburn, between Antelope Creek and the rim of Grand Canyon, honored the same friend — Assistant Secretary A. Bell — whose name appeared on "Bell's Peak" before it became Antler Peak. "Butler's Lookout" on the rim of Sheepeater Canyon, north of Osprey Falls, cannot be connected with a particular individual, but "Hibbard's Pass," the place where the original Norris Road crossed over from Obsidian Creek to Solfatara Creek, near Lake of the Woods, was named for Ed Hibbard, a local miner and part-time employee. "Johnston Peak," now Paintpot Hill, was named for employee N. D. Johnston.

"Longfellow Pond," one of the smaller lakes near Trout Lake — it could have been either present Buck or Shrimp lake — was probably intended to honor poet Henry Wadsworth Longfellow (a guess based on Norris's obsessive interest in poetry); "Mount Longfellow," now Mount Hornaday, a summit between Slough and Pebble creeks, is similarly accounted for. "Phelp's Peak," a presently unnamed summit east of Mount Washburn, was named for employee G. W. Phelps, whose homestead north of Gardiner was on a stream now called Phelps Creek. "Rowland Pass," the notch by which the trail around the east flank of Mount Washburn passed over the spur of the mountain, was named for another employee, George Rowland.

"Thompson's Peak," present Cook Peak, is a name which probably originated with Norris, since it appeared first on that feature on the map accompanying his annual report for 1880; however, the name had been shown, the previous year, on a minor eminence between Gardner River and Lava Creek — the one now marked 7717 on the 1961 Park-wide map. Just who Norris, was honoring remains in doubt. It is rather unlikely that it was Frank Thompson of the Northern Pacific Railway, as one writer has suggested on the basis of the map accompanying the 1881 *Report*, for the line was under construction in Dakota Territory in 1879, with the end of track only 125 miles west of Bismarck at year's end. Also, there was, at that time, no evidence of railway interest in Yellowstone Park, and when there was, Norris was quick to perceive that he had no friends among the Northern Pacific's managers — certainly none he would have honored with a place on the map.

One more name will complete the list of defunct Norris place names dedicated to people. "Toppin's Point," the first bulge in the shoreline of Yellowstone Lake above the outlet, on the west side, was named for Eugene S. Topping, whose residence there was noted by the Belknap party on August 1, 1875: "One Commodore Topping has established himself on the banks of the Yellowstone in a log hut and has a fine boat for the use of tourists."[42] Topping is the same man who, with Dwight Woodruff, discovered the Norris Geyser Basin in 1872, and later chronicled the early days of the upper Yellowstone country.

It was stated at the beginning of this section that two of the assistant superintendents (police under the civilian superintendents), George L. Henderson and Josiah W. Weimer, were responsible for many of the Park's place names. Henderson's contribution to the geographic nomenclature will appear in Chapter 7, since it was his involvement in the hotel business which led him to name features. Weimer named only five features, according to the scanty record available now, but his choice for four was superb; they have come down to us only slightly altered.

Four of Weimer's names were given to features on the terraces at Mammoth Hot Springs in 1883, the year of his appointment as an assistant superintendent. His "Bath Lake," a pool heated by nature to exactly the right temperature for pleasant bathing is now Old Bath Lake, an unfortunate decision of the Park's place name committee in 1927 when Bath Lake went dry (as it has in some seasons since). This name change developed into a troublesome duplication when the United States Geographic Board approved that name for a feature in the Lower Geyser Basin — the "Tank Spring" of the Hague Atlas (1904, Sheet xxii). In the early days of the Park, with dusty roads and plumbing yet a novelty, the soothing waters of Bath Lake were very popular — with men, that is. The pool was a masculine preserve for many years, and Henderson's *Guide* (1885) noted that "Ladies cannot even see it, for in pleasant weather large numbers of nude male bathers occupy it exclusively."[43]

The problem of male domination of "Great Bath Lake" (the earlier name) became a problem almost as soon as C. T. Hobart, manager of the Yellowstone Park Improvement Company, removed the dead trees and snags which had rendered Bath Lake dangerous before 1883. Bathing parties including ladies — they got as sweaty and grimy as any man while traveling the Park's dusty roads — had to send a man ahead to see if the swimming hole was a decent place for women; and that caused someone to suggest to the Livingston newspaper that "Ladies Lake," a smaller pond separated from Bath Lake by a ridge of limestone, ought to be "enclosed and kept exclusively for ladies."[44] However, Ladies Lake was so small and unsatisfactory that such a proposal was no better than a bad joke. A later suggestion that the government ought to regulate use of Bath Lake, perhaps build bath houses and require the use of bathing togs, was acted upon in 1912 with the building of a small bathhouse and requirement that all bathers be "properly dressed." Construction of a public swimming pool near Mammoth Lodge in 1925 ended the usefulness of Bath Lake, and the old bathhouse was removed three years later. National Park Service regulations no longer allow bathing there. It must be noted that Bath Lake is a name that did not satisfy everyone; W. H. Dudley (1885) suggested "Devil's Soup Bowl."

Weimer also named a nearby feature "Orange Geyser." While its output probably never qualified for geyser status, the color was right. Warm water wetting the sides of the ten-foot cone nourished algae which gave the formation its distinctive color. Orange Spring Mound, the approved name since 1927 (it has also been called "Grotto in the Glen," "Orange Geyser Cone" and "Orange Spring"). It is not a free-standing cone or mound but only the end of a low ridge of hot spring deposit. The buttressing ridge was used by some early visitors to ride horseback onto the mound to be photographed.

Another feature on the Mammoth terraces named by Weimer was "Cupid's Cave." The grotto was colored with algae fed with hot water through a fissure from "Cleopatra's Bowl" above. This lovely trysting place was romantically enhanced by Henderson, whose imagination named the entrance for Mark Anthony and found a horseshoe-shaped formation within — the "Gluck Auf," or good luck stone. By the time the United States Geological Survey mapped the terraces for Hague's Atlas, just after the turn of the century, the cave had been filled or closed over by hot spring deposit, and only an insignificant feature known as Cupid Spring remained after 1904. But the reason for the earlier name is explained by an old photograph which shows, in the head of a shallow gully, an opening probably sufficient for a person to enter by stooping.

The name of the feature above Cupid's Cave — "Cleopatra's Bowl" — may not have been entirely original with Weimer. The Hayden Survey is credited with naming the Cleopatra Spring which is still a recognized feature lower down on the terraces, and Weimer's usage could have been only a variant of that; regardless, it must have been Weimer's "bowl" which elicited Rudyard Kipling's wry comment concerning a pretty little basin "which some lurid hotel keeper had christened Cleopatra's Pitcher or Mark Anthony's Whiskey Jug, or something equally poetical."[45]

Weimer named at least one feature on the road to Norris Geyser Basin in 1883. His "Devil's Frying Pan," toned down to a mere Frying Pan Spring by the United States Geological Survey (1912), is another apt usage. The name came from the appearance given by gas bubbles escaping through a film of water, looking like grease popping on a hot griddle. Though the water was cold, the illusion remained, and stage coach drivers made the most of it by giving their passengers a line about how the birds in that locality drank so much hot water they all laid hard-boiled eggs.

If it seems like the men got to name all the Park features, that is not true. Mrs. E. N. McGowan, the wife of Assistant Superintendent Charles McGowan named Morning Glory Pool in 1883. She used, "Convolutus," a Latin name suggestive of the shape of the morning glory flower,[46] for the deep pool which has both the shape of the morning glory and its beautiful blue coloration. The name appears to have stabilized in its present form prior to 1889, for Mary Reeves, writing to her father from the Upper Geyser Basin hotel on July 30, used it so. Her description of that pool remains particularly good: "The deepest part of the pool was indigo blue, then it widens out in exact shape of the flower & of a lighter blue. The edge of corolla was crinkled & gray in color & the water bubbled over the edge."[47]

As a young ranger in the Old Faithful area in the 1930s, I remember the public's penchant for throwing coins into Morning Glory

Pool. Though those were depression days, there seemed to be plenty of people with pennies, nickels and dimes to throw into that beautiful pool — for luck, I suppose. One of my chores on Monday morning was to remove the coins and tax tokens from the walls of the pool, which was done with a little net on the end of a long bamboo pole. The gleaning was sent to the Park fiscal office to be remitted to the United States Treasury as miscellaneous receipts, though many of the coins were so eroded by the slightly acid water as to be of no further use as money.

An interesting alternative name suggested for Morning Glory Pool in 1891 was "Circe's Boudoir," but, if the enchantress of Homer's *Odyssey* had anything to do with that pool, it likely was to use it as a midden (an opinion based on the results of a 1949 pool cleaning project which yielded two truckloads of trash, including beer bottles, tin cans, hub caps and a wheel from some early automobile).

As mentioned in the introductory statement, the influence of officialdom on the naming of Park features was minimal during the Army period (1886 to 1917). One name which has survived from the time when Captain George S. Anderson was acting superintendent illustrates the camaraderie existing in the cavalry regiments of the "Old Army." The captain enjoyed an occasional backcountry patrol, and, on such a trip into the Falls River Basin in 1896, the small detachment stopped for the night on a stream which had no name. As camp was being made, Captain Anderson said to the men, "I'll make you a proposition — you get the water and I'll furnish the whiskey," and one of the men replied, "Captain, we'll just name this creek for your proposition!"[48] It has been called Proposition Creek ever since.

Captain John Pitcher, who was the Acting Superintendent when Old Faithful Inn was built, has the honor of being the first, of record, to use that name for the Park's famous hotel (in his annual report for 1904), though that does not mean he originated the name.

Officers of the army Corps of Engineers were responsible for the construction and maintenance of roads and bridges in the Park from 1883 to 1918, and yet, they had surprisingly little to do with naming its features. Major Charles J. Allen, who had charge of the work during the seasons of 1888 and 1889, may have named Arnica Creek, which made its debut on his map showing the alignment for a proposed road from the outlet of Yellowstone Lake to the Upper Geyser Basin by way of the south shore of Shoshone Lake.[49] The same map shows a "Lookout Point" on the rim of the Grand Canyon east of

Mount Washburn (reminiscent of the "Paddy's Lookout" name which appeared there in 1881).

Captain Hiram M. Chittenden, the Park's great engineer and historian, has already been mentioned as responsible for several important place names. To those can be added Craig Pass, "From the maiden name of Mrs. Ida Craig Wilcox, the first tourist to cross the pass."[50] While most later editions of Chittenden's book imply that the Isa Lake name — concocted from the given name of Isabel Jelke — was also a product of that excursion, it is not so; the young lady was a Northern Pacific Railroad tourist of 1893 (cf. 1895 edition, p. 331).

One new name appearing on the Jones map of 1894 — Terraced Falls — may have originated with Chittenden since he provided the information for that map,[51] and four place names which appear for the first time in his book — Firehole Falls, The Hoodoos [limestone], Lewis Falls and Pelican Valley — can be attributed to him until proven otherwise.

Two engineer officers whose ideas contributed most to the development of Yellowstone's system of park-type roads are now honored by place names. Lieutenant Dan C. Kingman, the man who got the road from Mammoth Hot Springs onto Swan Lake Flat by constructing a timbered trestle around the wall of Glen Creek canyon (the Golden Gate), and thereby avoided the difficult Snow Pass route used by the "Norris Road," is remembered by Kingman Pass where his new road reached the meadows which surround Swan Lake. It was Kingman who envisioned the present interior loop road which has served the Park so well; he turned the primitive wagon trails he found in 1883 into roads as good as any in the West.

Dan Kingman's trestled roadway through the Golden Gate was replaced in 1900 by a one-lane concrete structure, which took the name of its designer, engineer Chittenden. This "Chittenden Viaduct," or "Golden Gate Viaduct," was widened to two lanes in 1934, and entirely rebuilt in 1977. An interesting feature of the wooden trestle was a rock pillar on the canyon side of the road at the lower end of the trestle. Upon construction of the original viaduct, the twenty-three-ton pillar — called "Imperial Rock" in 1884[52] — was broken from its natural base and moved out and down the road equal distances to a new foundation. It was moved again at the time of each reconstruction of the viaduct.

Captain Hiram M. Chittenden, who pushed the road system to a magnificent completion and thus set the standard for park-type roads for years to come, has been honored in the names given two of his other creations. The Melen-arch bridge he constructed in 1903 to span the Yellowstone River above its upper fall was called Chittenden Bridge from 1912 until it was replaced by a structure better-suited to modern traffic needs in 1963. The new structure was dedicated as the Chittenden Memorial Bridge. The other work which bears Chittenden's name is the road up Mount Washburn, which was so designated in this manner:

> The Secretary of the Interior, Hon. Franklin K. Lane, and the assistant to the Secretary, Mr. Adolph C. Miller, visited the Yellowstone National Park from July 30 to August 3, 1913. They were impressed with the excellence of the road and bridge system laid out by Gen. Hiram M. Chittenden while engineer officer in charge of the work, and especially with the road from the canyon to the top of Mount Washburn. The Secretary decided that the valuable work of Gen. Chittenden was worthy of recognition, and, on motion of Mr. Miller, announced that hereafter the road from the Canyon Junction to the top of Mount Washburn be known as 'Chittenden Road,' and directed the acting superintendent to make that fact of record and cause suitable sign posts to be erected.[53]

Early in the period of civilian administration which followed establishment of National Park Service oversight in 1916, a place name committee of three (consisting of Park Superintendent Horace M. Albright, photographic concessioner Jack E. Haynes, and ranger-naturalist Marguerite L. Arnold) was created to handle the problems involved in the naming of Park features and revision of unappropriate old usages. Much that was done by the committee is now difficult to trace, and exactly who was responsible for what is often less certain than it ought to be. However, it does appear that Mr. Albright was most interested in names drawn from people (he was a lawyer by training), while Jack Haynes, whose schooling was in geology, concentrated on thermal geology. Just what Mrs. Arnold's contribution was is unknown.

Among the place names which can be attributed to Mr. Albright is Clagett Butte (1926) for William H. Clagett, a Republican Delegate representing Montana Territory in the Forty-second Congress (March 4, 1871, to March 3, 1873). Though he later claimed credit for drawing the bill which became Yellowstone's organic act, and for doing, with Langford, "two-thirds, if not three-fourths of all the work connected with its passage,"[54] that is not in accord with the facts. A discussion of the events which led to the passage of the Yellowstone legislation, appears in my book, *The Yellowstone Story* (1977), Vol. I, pp. 165–172. It is sufficient to say that the drafting of the bill was not an original work (it was drawn on the model of the 1864 Yosemite Grant legislation of Senator Dawes), and there is no evidence that Clagett did anything beyond introduction of the House version of the bill (it was the Senate version — Samuel C. Pomeroy's S. 392 — which passed both houses and was signed into law by President Ulysses S. Grant on March 1, 1872).

It would seem that Senator Pomeroy, whose twelve years in the Congress, during which he served as chairman of the Senate Committee on Public Lands and was always a friend of the then new Park idea, deserved recognition on the Park map ahead of Clagett, whose only service to the Park and what it represented lay in his introduction of a companion measure, H.R. 764, of no ultimate consequence. Nor was there the stuff of greatness anywhere else in Clagett's career; he was, throughout, the type of restless, opportunistic politician so common in the West during the years when the territories and states were taking shape. In Nevada, Montana, Colorado, the Black Hills of Dakota, and Idaho he was interested in frontier politics and mining rather than the legal practice he was so thoroughly trained for, and those interests always played him false.

Mr. Albright was on safer ground in putting the name of Charles W. Cook on that peak in the Washburn Range first called "Thompson's Pk." by Superintendent Norris (1880), and "Storm Peak" by the United States Geological Survey (1885). It was an excellent decision since it put Cook's name adjacent to the peak which immortalizes David E. Folsom, his comrade on that bold exploration of the Yellowstone Region in 1869; but beyond that, it was nice that the recognition came in 1922, while Cook was yet alive. He lived on, to January 30, 1927, the last of the early explorers, lionized at the celebration of the Park's semicentennial year and the recipient two years later, of a letter on White House stationery which read:

February 14, 1924

My dear Mr. Cook:
Through the courtesy of Mr. Cornelius Hedges, Jr., and of Congressman Scott Leavitt, I have learned that you are within a few days to celebrate your 85th birthday anniversary. As one of the group of men responsible for the founding of the national park system, you have rendered a series of national services of truly notable character. Upon these I wish to extend my felicitations, and my congratulations upon your approaching birthday. I hope you may live to enjoy many more celebrations of the same anniversary.

Most Sincerely yours,
Calvin Coolidge[55]

Another of Superintendent Albright's place names, Phillips Caldron, in the Norris Geyser Basin, remembers Ranger-naturalist Charles Phillips who died tragically at Old Faithful on April 11, 1927. Phillips was stationed there during the winter of 1926–1927 to make wildlife and geyser observations; an assignment which was not considered particularly dangerous since rangers, and the soldiers who were the Park's protectors before them, had stayed in the geyser basins overwinter for many years. Also, he was not alone as the hotel winter keepers, Mr. and Mrs. Bauer, lived nearby.

On the evening of April 11, Mr. Bauer brought Phillips some plants he had found growing in a warm rivulet near his house and asked the ranger if he could identify them. The decision was that they were a species of wild parsnip which has an edible root, whereupon they decided to try some for dinner. The meal was eaten at Bauer's house about 8 o'clock, and Phillips went back to the ranger station at 11:15. Both the winter keeper and his wife were taken violently ill about two in the morning, suffering waves of nausea and vomiting which made them so weak they stayed indoors until the evening of the following day. Since they had heard nothing from Phillips, Mr. Bauer put on his "webs" and crossed the several hundred yards of snow to the ranger station — where he found the building dark and cold, and Phillips dead on the kitchen floor. The plants they had eaten were a variety of the deadly water hemlock.[56]

President Warren G. Harding's visit to Yellowstone Park on June 30 and July 1, 1923, coincided with the appearance of a new geyser in the Norris basin, and Superintendent Albright thought it would be appropriate to put his name on the spouter. But the president died in office a month later, on August 2nd, and the geyser didn't last out the summer either, so that the name had to be changed to Harding Geyser Crater.

Two more of Albright's place names honoring persons important in the Park's history — Howard Eaton Trail, for the initiator of the "dude ranch" business in the Dakota-Wyoming-Montana area, and Mount Haynes, for the pioneer photographer and concessioner, Frank J. Haynes (1853–1921) — are discussed in Chapter 8.

Superintendent Roger W. Toll was responsible for three place names of lasting interest: Lemonade Lake (1934), the pond of greenish water which formed a pool beside the parking area at the foot of Roaring Mountain — now dry, and Rainey Lake (1932), a pond between Tower Junction and Overhanging Cliff, where gas bubbles rising from below create the illusion of raindrops spattering on the surface. The third place name was Mount Jackson, for the Park's pioneer photographer, William Henry Jackson (1843–1942). Appropriately, his memorial is on the north side of the Madison River, across from Mount Haynes.

## NOTES

1. G. L. Henderson, *Yellowstone Park Manual and Guide* (Mammoth Hot Springs, Yellowstone National Park, July 1, 1885), p. 4.
2. Hiram M. Chittenden, *The Yellowstone National Park*, ed. by Richard Bartlett (Norman: University of Oklahoma Press, 1964), quotation on pp. 157–158.
3. "The Yellowstone Park — Its Management," Bozeman, MT, *Avant Courier*, September 30, 1880.
4. P. W. Norris, *Report Upon the Yellowstone National Park to the Secretary of the Interior... For the Year 1878.* (Washington: Government Printing Office, 1879), p. 6 & map.
5. P. W. Norris, *Report Upon the Yellowstone National Park to the Secretary of the Interior... For the Year 1879* (Washington: Government Printing Office, 1880), pp. 5, 16, 18, 25, 28 & map.
6. P. W. Norris, *Fifth Annual Report of the Superintendent of the Yellowstone National Park...* (Washington: Government Printing Office, 1881), pp. 8, 22, 54, 55, 62, 64 & map.
7. Norris, *Report* for 1879, op. cit., p. 25 & map.
8. Probably in 1878, since the topographic map (*Yellowstone National Park*, 1:126,720) is identified as surveyed in that year. As noted, this map did not appear in published form until after 1881.

9. M.M. Quaife, ed., *Yellowstone Kelly* (Lincoln: University of Nebraska Press, 1973), p. 228.
10. Norris, *Report* for 1879, op. cit., p. 6.
11. P. W. Norris, *The Calumet of the Coteau . . . Together With a Guide-Book of the Yellowstone Park* (Philadelphia: J. B. Lippincott & Co., 1884), p. 262.
12. Hiram M. Chittenden, *The Yellowstone National Park,* ed. by Richard Bartlett (Norman: University of Oklahoma Press, 1964), p. 158. The comment on p. 150, concerning Dunraven Peak, is not accurate.
13. C. W. Cook, D. E. Folsom and Wm. Peterson, *The Valley of the Upper Yellowstone . . . 1869,* ed. by A.L. Haines (Norman: University of Oklahoma Press, 1965), p. 43.
14. The 1878 map; see Note 8.
15. P. W. Norris, *Fifth Annual Report* (1881), p. 60; see Note 6 for citation.
16. Ibid., p. 62.
17. Ibid.
18. P. H. Sheridan, J. F. Gregory and W. H. Forwood, *Report of an Exploration of Parts of Wyoming, Idaho and Montana, in August and September, 1882* (Washington: Government Printing Office, 1882, p. 28. The naming of the geyser is mentioned in the Bozeman, MT, *Avant Courier,* September 15, 1881.
19. From *The Chicago Tribune,* December 6, 1891.
20. United States Geological Survey of the Territories, *Map of the Egeria Springs, Firehole River, . . . (1878).*
21. P. W. Norris, *Calumet of the Coteau* (1884), p. 252; see Note 11 for citation.
22. P. W. Norris, *Fifth Annual Report* (1881), p. 69; see Note 6 for citation.
23. Aubrey L. Haines, "Visit to the Queen's Laundry Bathhouse," typed report to the Chief Park Naturalist, June 14, 1961, p. 3. Yellowstone Park Reference Library.
24. Hiram M. Chittenden, *Yellowstone National Park,* 4th ed. (Cincinnati: The Robert Clarke Company, 1903), pp. 131–132.
25. National Park Service, *Yellowstone National Park, Wyoming, 1923* (Washington: Government Printing Office, 1923), p. 43.
26. United States Geological Survey, *Norris Junction, Wyo.,* N4430-W11030/15 (1958).
27. George Catlin, *Letters and Notes on the Manners, Customs, and Conditions of the North American Indians,* 2 Vols. (New York, 1841), I, pp. 261–262.
28. Josiah Whitney, *The Yosemite Book* (1868), p. 22; quoted by Albert Matthews, "The Word Park in the United States," *Publications of the Colonial Society of Massachusetts,* VIII (April 1904), p. 385.
29. "Our National Park," *Helena* (MT) *Daily Herald,* February 28, 1872.
30. N.P. Langford, *Diary of the Washburn Expedition to the Yellowstone and Firehole Rivers in the Year 1870* (St. Paul: F. J. Haynes, 1905), pp. xxiii–xxiv. This account, which is alternatively titled *The Discovery of Yellowstone Park, 1870,* is not truly a diary, but only a recollection which is reinforced from diary sources.
31. Aubrey L. Haines, *The Yellowstone Story,* 2 Vols. (Colorado Associated University Press: Boulder, 1977), I, pp. 89–90, 100–103. Other aspects of the origin of the "national park idea" are considered on pp. 137–140, 155–173.
32. Ibid., p. 155.
33. Jack E. Haynes, *Haynes Guide, Handbook of Yellowstone National Park* (St. Paul, MN: Haynes, Inc., 1939), pp. 58–59.

34. Jack E. Haynes, *Haynes Guide, Handbook of Yellowstone National Park* (Bozeman, MT: Haynes Studios, Inc., 1961), p. 110.
35. Roger W. Toll, "Annual Report for Yellowstone National Park, 1933" (mimeographed), p. 3.
36. William R. Bandy, "Plat of Survey of the East Boundary of Yellowstone National Park," September 27, 1931.
37. Joseph Joffe, "The Grand Old Man of Yellowstone National Park, John W. Meldrum," typescript, 1940. Yellowstone Park Reference Library, 131 pp.
38. *United States, Statutes at Large*, Vol. 28, p. 73.
39. Philetus W. Norris, *Annual Report of the Superintendent of the Yellowstone National Park. . .1880* (Washington: Government Printing Office, 1881), p. 5. An account of this visit, with sketches of features, follows on pages 6 through 8.
40. "There are fifty to sixty persons . . . ," Bozeman, MT, *Avant-Courier*, Aug. 4, 1881, p. 2.
41. P. W. Norris, *Calumet of the Coteau* (1884), p. 262; see Note 11 for citation.
42. W. E. Strong, *A Trip to the Yellowstone National Park in July, August and September 1875* (Washington: privately printed, 1876), p. 80.
43. G. L. Henderson, *Yellowstone Park Manual (*1885), p. 1; see Note 1 for citation.
44. "Park Paragraphs," *Livingston* (MT) *Daily Enterprise*, November 21, 1883.
45. Rudyard Kipling, *From Sea to Sea* (Garden City, New York: Doubleday, Page & Co., 1920), p. 74.
46. Information from Mrs. William Starr Brackett, daughter of Mrs. McGowan, November 9, 1950. Yellowstone Park Reference Library.
47. Mary Reeves, p. 4. Yellowstone Park Reference Library.
48. C. Max Bauer, "Place Names of Yellowstone National Park," 1937, p. 53. MS, Yellowstone Park Reference Library.
49. Charles J. Allen, "Map of the Yellowstone National Park, Showing Roads Completed, Under Construction, and Approximately Projected, November 14, 1888." Yellowstone Park Reference Library.
50. H. M. Chittenden, *Yellowstone* (1964), p. 179; see Note 2 for citation.
51. William A. Jones, "Road Map" [Yellowstone National Park], 1894. Yellowstone Park Reference Library.
52. Thomas Henry Thomas, "Yellowstone Park Illustrated," *The Graphic*, August 11, 1888; I, p. 160. This two-part article covered a Park trip in September and October, 1884.
53. Lloyd M. Brett, *Report of the Acting Superintendent of the Yellowstone National Park . . . 1913* (Washington: Government Printing Office, 1913), p. 9.
54. N. P. Langford, *Diary* (1905), p. xxiii; see Note 30 for citation.
55. Letter, President Calvin Coolidge to C. W. Cook, February 14, 1924. Original held by Cook's daughter, Mrs. Josephine Mueller, Lewistown, Montana.
56. A. L. Haines, *The Yellowstone Story* (1977), II, pp. 314–315; see Note 31 for citation.

# 7

# CONCERNING MINE HOST

| | |
|---|---|
| *Mammoth* | *McCartney* |
| *Cottage* | *Black Growler* |
| *Sapphire* | *Marshall* |
| *Yancey* | *Uncle Tom* |
| *Larry* | *Miscellany* |

The place names traceable to concessioners operating public facilities in Yellowstone Park and their employees show no distinct patterns but have appeared randomly throughout the span of white occupation — usually without generating spin-offs, those minor usages that tend to develop out of important place names. The names treated here have much in common with those to be considered in the next chapter (the legacy of the visitors); both are mainly accidental accretions.

There were innkeepers in the Yellowstone region before it was a park. An informal gathering of invalids early in the summer of 1871 — around the outflow of hot water entering Gardner River below Mammoth Hot Springs — soon became "McGuirk's Medicinal Springs," with Matthew McGuirk as its proprietor. At nearly the same time two Bozeman men, Henry R. Horr and James C. McCartney, laid claim to the hot spring terraces above McGuirk. Henry soon sold his part in the claim to Jim, who turned the cabin they had built at the mouth of Clematis Gulch into a rude hotel.

Neither McGuirk nor McCartney filed on their land claims before the passage of the Yellowstone Park Act, March 1, 1872; therefore, their claims were invalid. They eventually received some compensation for improvements taken from them by the government, but they did not gain title to the land. McGuirk was evicted by Superintendent Langford in August of 1874, and McCartney was dispossessed by Superintendent Norris in 1881 (reluctantly relocating, as we have seen, to the vicinity of present Gardiner, Montana, where he settled on a land claim straddling the Park boundary).

Another innkeeper got a foothold in the Park during the summer of 1880 when George Marshall built a log house in the Lower Geyser Basin described as "a fine shingle-roofed mail station and hotel" to serve traffic on the stage line being established to connect the Park with Virginia City. The "Virginia City and Hot Springs Line" was a failure, but Marshall hung on and parleyed his improvements into a two-story hotel — the "Marshall House."

As construction of the Northern Pacific Railroad approached the Park, its management recognized the tourist potential of the area and took an interest in tourist facilities; in fact, they decided to get control of the Park business. The plan involved leasing of land surrounding all the principal features by a dummy concern, the "Yellowstone Park Improvement Company," which could serve the railroad's tourists unrestrained by competition. This was arranged through collusion in the office of the secretary of Interior, but the inherent monopoly it would have created was recognized by Senator George Graham Vest, who introduced legislation to limit the railroad's venture.

The Yellowstone Park Improvement Company began construction of facilities in the Park in 1882, and had nearly completed a large hotel at Mammoth Hot Springs by the following August when the Northern Pacific's "Park Branch Line" was completed to its Cinnabar terminus three miles from Gardiner, Montana. The large number of visitors who inundated the Park following establishment of rail transportation had to make do in the unfinished hotel at Mammoth Hot Springs and in tent camps at Norris, Old Faithful, and Canyon.

Badly mismanaged from its inception, the Yellowstone Park Improvement Company was bankrupt by the beginning of the 1884 season, and its assets were sold at sheriff's sale on the order of a Wyoming court in the fall of 1885. The buyer was the Northern Pacific Railroad, which acted to protect its interest. The business was reorganized as the Yellowstone Park Association, which the railroad always controlled, even though its financial relationship — often disguised under a "Northwest Improvement Company" cover — varied greatly over the years.

The Yellowstone Park Association proceeded to construct permanent hotels at five other locations in addition to Mammoth Hot Springs. Following is a brief history of these structures:

- *Old Faithful.* Originally built in 1885, this hotel was destroyed by fire in 1894 and replaced by a temporary structure that served until Old Faithful Inn was constructed in 1903.
- *Canyon.* A small, prefabricated building was erected in 1886 and served until 1890, when a permanent hotel was built. It was replaced in 1911 by a large hotel abandoned in 1959 — and razed in 1960.
- *Norris.* A fine hotel built in 1886 was destroyed by accidental fire in the spring of 1887, as it was being opened for the summer season. Another hotel was built on a different site in 1901 and abandoned in 1917.
- *Fountain.* A fine hotel was built in 1890 and abandoned in 1917, when motorization of the Park's public transportation made it and Norris unnecessary.
- *Lake.* Built in 1890 and in use to the present, with additions and changes, Lake Hotel is now the Park's oldest, as none of Mammoth's old "National Hotel" survived the altering hand of successive managements.

Meanwhile, competition developed to plague the Association. George L. Henderson, one of the early assistant superintendents, or Park police, was the brother of David Henderson, Speaker of the House of Representatives, and that relationship seems to have sufficed to get a hotel concession for the Henderson family in 1885. Their "Cottage Hotel Association," organized by son Walter and daughter Helen, built a log hotel at Mammoth Hot Springs, and also bought the Marshall House in the Lower Geyser Basin, using both to undercut the shaky Yellowstone Park Association. As a further problem, an illegitimate hotel appeared at Old Faithful at that time (the product of collusion between Superintendent Robert E. Carpenter and the brother of the incompetent manager of the Yellowstone Park Improvement Company).

The Yellowstone Park Association bought out the Hendersons in 1889 and having established its legal right to the hotel at Old Faithful was finally in position to monopolize the Park tourist business. Just as things seemed to be going their way, with new hotels under construction at Grand Canyon and the Lower Geyser Basin in 1890 and at Lake in 1891, there was trouble in the form of loss of the transportation privilege (always the profitable side of the tourist business). With transportation in the hands of a separate company, the Association's hotel business was left dependent upon railroad subsidies to make up the annual operating deficits.

Inroads on the business of both the hotel and transportation companies began in 1893, when William Wallace Wylie, a Bozeman schoolteacher, obtained a permit allowing him to operate a low-cost tourist business based on permanent tent camps. This "Wylie Way" rapidly became popular, leading to a further dilution of the total Park business as other upstart companies obtained a foothold. By the time the new National Park Service took over management of Yellowstone Park in 1916, its concession business was chaotic. Director Stephen T. Mather, who left a successful private business to shepherd the nation's parks, believed in controlled monopoly, and he set about correcting the situation in Yellowstone Park by consolidating the services offered visitors.

The reorganization effected in 1917 brought three separate enterprises — the hotel, transportation, and boat franchises — under one management, headed by Harry W. Child (who agreed to motorize all Park transportation). It also combined the several camping companies into a Yellowstone Park Camping Company, without transportation privileges, leaving that business to reorient toward the automobilist. In 1928, Mr. Child bought the Camping Company, so that his enterprise consisted of an amalgam of five businesses: hotel, transportation, boating, fuel, and lodges (this last created out of the permanent camps operation). It only remained for Child's successor, William M. Nichols, to merge those elements — in 1936 — into the Yellowstone Park Company, which has passed through several management changes to the present day. A similar process has acted to consolidate merchandising — the store and picture shop businesses — under one management: Hamilton Stores. Thus, the Park's visitation is now served by two concessioners.

It would be logical to expect that such varied business enterprises, extending over so long a span of time and employing so many people, would result in the appearance of some important place names down through the years — but that has not been the case since the pioneering period. And of those names that are our legacy from the Park's commercial past, many — but not all — have come from the pen of one man — George L. Henderson — called a "lurid hotel keeper" by Rudyard Kipling (who read his newspaper-like guide to the Park and its features). George L. Henderson's penchant for writing made him godfather to a host of names, though relatively few have survived.

## MAMMOTH

This usage, as it appears in the name of Mammoth Hot Springs, is the most important place name related to the Park's concession business, and also the oldest from that source. The name was given Yellowstone's largest hot spring area by Harry R. Horr when he and James C. McCartney both claimed 160-acre tracts, including the hot spring terraces, on July 5, 1871.[1]

From a letter by "Semi Occasional," appearing in the Helena *Rocky Mountain Weekly Gazette* of July 24, 1871, it is evident that Horr and McCartney were in the process of erecting log houses in the mouth of Clematis Gulch on June 24, but the writer still used the older designation of "White Sulphur mountain" in referring to the hot springs.[2]

By September, the hot springs Horr and McCartney claimed had a new name — Mammoth Hot Springs. The editor of the newspaper at Bozeman, Montana, had the name directly from Harry Horr, who had stopped at the office for a chat. In addition to recording the name in his paper, the editor also noted: "There are ten patients now at the springs, and all are doing well. We congratulate Messrs. Horr and McCartney on their success in getting what will at an early day be a big thing."[3]

The spa Horr and McCartney established at Mammoth Hot Springs was not a success, but the name was just right. It avoided the mistakes inherent in the use of "sulphur" and "soda" — the dissolved and redeposited limestone of the formation was neither, chemically speaking. Also, the new name accented the great size of the terraces (preeminent in all the world after their New Zealand rivals were destroyed by a volcanic eruption in 1886).

For those who might wish some certain proof that Harry Horr was the author of the name Mammoth Hot Springs, there is his own statement: "I gave the Springs the name they now bear."[4]

Something needs to be said as to why the Horr and McCartney enterprise was an eventual failure. The passage of the Organic Act creating Yellowstone National Park on March 1, 1872, took it out of the public domain, and also created confusion as to what possessory rights, if any, might remain to those persons who had settled within the bounds in good faith. As mentioned, Harry Horr and James McCartney had declared their intentions by filing at the Gallatin County Courthouse (Bozeman, MT) on July 5, 1871, and in Uinta County (Evanston, W.T.) in the following month, while Matthew McGuirk did not file on his claim at Hot River until March 9, 1872 — eight days after the Yellowstone act was signed into law. In the case of Horr and McCartney, the fact that the land claimed had not been opened to settlement left them in the position of squatters, with no more assurance in their tenancy than a priority in claiming the tracts — if and when the district was opened to settlement. McGuirk, whose claim was not made until *after* Yellowstone Park was withdrawn from the public domain, did not have even the tenuous right of a squatter. After passage of the Yellowstone Act, Harry Horr made a feeble effort to confirm his claim, but after a few inquiries in 1873 either sold to his partner or abandoned the claim to him. Jim McCartney stayed on at the Springs, where he ran a rude hotel on a sporadic basis until evicted by Superintendent Norris in 1881.

This hostelry — where the creature comforts were limited to just enough floor space for the guest to put his blankets down and a monotonously plain "grub pile" — was variously known during the decade of its operation as the "Pioneer Hotel," "Job's Hotel," and "McCartney's Hotel," but the last was most common.

There have been some spin-offs from the popular Mammoth place name. There was a futile attempt to fix the name of "Mammoth City" on the new town site developing out of McCartney's ranch at the North Entrance to the Park in 1883 (instead, it became Gardiner, Montana). That same year, a feature on the upper terraces was called "Mammoth Orange Geyser," although it probably never qualified as a geyser. It is now Orange Spring Mound. George Henderson came up with "Mammoth Cave" (1885) for a no-longer-identifiable feature below the White Elephant Back Terrace. The structure built as the "National Hotel" in 1882–1883 and known for years by that name became simply "Mammoth Hotel" in 1903, and "Mammoth Hot Springs Hotel" the following year. Still later, as the carriage trade declined, the hotel metamorphosed into "Mammoth Motor Inn." There was also a "Mammoth Camp" — the lodge development of the 1920s, below Jupiter Terrace — and the whole headquarters settlement became "Mammoth Village" after the facelift that gave it the landscaped "Esplanade" in the mid-1930s.

There is no relationship between the usages just mentioned and several that once attached the word "Mammoth" to thermal features at Norris and in the Lower Geyser Basin.

## MCCARTNEY

Many people have been immortalized through the attachment of their names to Park features, but only one — James C. McCartney, the pioneer hotel man — is remembered on the map by a hole in the ground!

In his day, McCartney's name was associated with the rude hotel he operated sporadically in the mouth of Clematis Gulch, and the stream of water flowing past his door was, of course, "McCartney Creek." Even the original road to Mammoth Hot Springs (a steep ascent over the hill rather than up Gardiner River) was known as "McCartney's Hill" because he built it. In addition to those, there was "McCartney's Cave," an inconspicuous solution hole in the Hotel Terrace. The hotel, which ceased to be McCartney's in 1881, burned in 1912 while in use as a Chinese laundry; the stream issuing from the gulch was renamed Clematis Creek by the Geological Survey in 1904; and the hill lost its significance and was forgotten after the road was relocated on a better grade. Only the cave remains under McCartney's name.

Historian Chittenden has explained the origin of the name, now properly applied as McCartney Cave, thus: "In the last mentioned cave, or, more properly, crater, an elk fell one winter when the crater was level full with light snow. His antlers caught between the sides of

McCartney's Cabin. This rude log cabin that stood in the draw near Hymen Terrace was the Park's first and only "hotel" until 1880.

the crater, holding him in a suspended position until he perished. He was found the following spring by Mr. McCartney."[5]

The date of the foregoing incident is unknown, though a 1902 visitor placed it in 1869 (an impossibility because Jim McCartney did not take up residence at Mammoth Hot Springs until two years later). The comment of a visitor of 1885 is probably typical of the feelings concerning this cave:

> Superintendent Erret showed me the openings of two caves on the flat near the hotel, one of which has been explored to the depth of 350 feet. Water can be heard below. . . . I asked Mr. Erret why these caves had not been explored to a greater extent. He said they had never yet found a man who hankered after travelling around down in the earth, upon those treacherous shelves . . . and that the Park put up at auction wouldn't bring enough to hire him to go down there. On second thought I agreed with him.[6]

About twenty-five years ago, I attempted to explore McCartney Cave but found the passage blocked by fallen debris less than a hundred feet from the entrance, which was just as well since carbon dioxide gas is known to accumulate in dangerous amounts in caves on the upper terraces and might be encountered there also. I am in complete agreement with the man whose considered estimate of spelunking under the terraces was that "the results would be small, a probable total of darkness, dirt, bad odors, and bruised extremities."[7]

## COTTAGE

Though built as the Cottage Hotel, the log building erected by the Hendersons at Mammoth Hot Springs in 1885 was an employee dormitory during most of the seventy-nine years of its existence.

While the earliest reference to the Cottage Hotel under that name appears in a newspaper-like guide published by George L. Henderson July 1, 1885, months before the building was actually completed, it would seem that the project had its beginning early in the year. A ten-year lease for a site on which to construct a hotel and some other buildings was granted Walter J. and Helen L. Henderson, son and daughter of Assistant Superintendent George L. Henderson, on March 3, 1885. The lease was probably obtained through the influence of David B. Henderson, Speaker of the House of Representatives and the brother of George. It is also likely that this thinly veiled maneuver, by which "that old Devil Henderson" — meaning George — got into the Yellowstone hotel business, was the real source of Superintendent Wear's great dissatisfaction with his number-one employee (a squabble ending in Henderson's resignation on June 2, 1885). Be that as it may, the Cottage Hotel project was progressing.

A plan for the building dated April 3, 1885, is in the Ash Collection at the Yellowstone Park Reference Library, and a news item of July 1 shows that George L. Henderson was, indeed, the prime mover:

> Mr. G. L. Henderson is about to begin the erection of a building 36 × 40 feet on his leased ground at Mammoth. . . . This is the beginning of a series of buildings to include a large hotel, a number of cottages and a stable, which he plans to erect for the accommodation of tourists in plainer style and at less cost than is in vogue at the National . . . [it] will be constructed of dressed, round logs, carefully laid and tastefully finished within and without. Gordon Bros., of Livingston will superintend.[8]

The three-story hotel was ready for plastering in November, and Henderson was in the East buying furnishings; at least that was the first report. However, it soon developed that the trip had had another purpose — an attempt to lease the new hotel to a syndicate representing the Northern Pacific Railroad.[9] But that group was determined to get control of the National Hotel instead, and finally did, at sheriff's sale.

It took the syndicate several years to gain control of all properties of the defunct Yellowstone Park Improvement Company and reorganize them as the Yellowstone Park Association, but in 1889 they were in position to buy out Henderson's Cottage Hotel Association, which by then had control of the Marshall House also. By that purchase, the hotel business in the Park was consolidated under one company controlled by the Northern Pacific Railroad.

From that time on, the Cottage Hotel was used as a dormitory for summer employees and for company management people during the winter season. The old building was razed on May 4, 1964, after standing vacant for several years.

## BLACK GROWLER

This usage is a peculiar case. It made an initial appearance as the "Black Geyser," but it remained for hotel man George L. Henderson to sense its true character and popularize it as the "Black Growler." Time has formalized that as Black Growler Steam Vent.

Dr. A. C. Peale of the Hayden Survey, who observed and named the Black Geyser in 1878, must have seen it at a moment of most unusual activity. Its character subsequently has been that of a superheated steam vent with a tendency to occasionally shift its activity to new openings. It was that continuous, noisy discharge of invisible steam that led to the change in usage from Black Geyser to Black Growler. Whether that came about through common usage or was Henderson's inspiration is not clear, but he did popularize the more accurate name. There is a hint of the variability of this thermal feature in his statement: "The Black Growler has been nearly dead but has been increasing in vigor during the last year."[10]

The *Haynes Guide* for 1896 contains a description indicating that this feature was still vacillating between geyser and steam vent:

> The Black Growler, with a chimney-shaped opening, is located quite near the road, at the head of a gulch leading from the plateau. Very little water is thrown out, while a large quantity of steam is constantly escaping, producing a peculiar sound. The deposit surrounding the edge of the crater is, at times, of inky blackness, evidently the origin of its name. The water is not clear and has a strong odor of sulphur, which is probably the cause of its turbidity. The geyser, a few feet to the north, known as the 'Hurricane,' is similar to the Black Growler in the character of its eruptions, which are very irregular, and destructive to surrounding vegetation and foliage.[11]

Through the years up to 1928, the Black Growler retained its guide book position as one of the preeminent features of Norris Geyser Basin. As it shifted increasingly toward activity as a superheated steam vent, there was the added interest of a visual demonstration. Its roaring column of invisible steam, escaping at a temperature of 284 degrees Fahrenheit (discernible, on a hot summer day, only by the distortion of objects viewed through it) could be made visible by blowing tobacco smoke toward the vent — dangerous, of course, but it did outline the escaping steam by the vapor condensing on the smoke particles. Another interesting demonstration consisted of exposing a stick in the orifice for a few minutes, which resulted in carbonizing the wood, leaving no doubt about the heat latent in the steam.

In 1928 the Black Growler shifted its activity to another opening lower down the slope, and after a few seasons there, it moved again to a vent on the other side of the walkway, where it performs insignificantly as Black Growler Steam Vent.[12]

## SAPPHIRE

The name of the superheated Sapphire Pool, like some other important usages attributed to George L. Henderson — Golden Gate, Hymen Terrace, Narrow Gauge Terrace, and Porcelain Basin — is credited to him on the shaky basis that he is the first one known to have used it.

This feature was noted as "Soda Geyser" in the Hayden Survey report for 1872,[13] because of its apparent effervescence when disturbed. That effect is a result of the pool's superheated condition,

Sapphire Pool. The name comes from George L. Henderson's characterization of it as "the great Sapphire Geyser." He called this area "Biscuit Terrace" (present Biscuit Basin) from the shape of the encrustation around the pool.

which occurs when the water is hotter than normal at that elevation, with accumulation of latent energy. All that is required to release energy in the form of steam is a lowering of surface tension. Thus, a vagrant breeze, stirring with a stick, or introduction of dust can create a disturbance varying from a minor effervescence to a massive upheaval through breaking of the invisible barrier preventing the rapid escape of steam. The same mechanism operates in several other pools but was not fully understood for a time; as late as 1896, the *Haynes Guide* called such pools "pulsating or breathing springs (geysers in fact)" (see p. 73).

Henderson was probably unaware of this deep pool's earlier name when he renamed it "the great Sapphire Geyser" (for the gem color of its water), because he writes like one standing on the threshold of a pristine place when he states: "There are many discoveries yet to be made in this most wonderful and fascinating basin."[14] True enough, it was difficult to reach from the tourist route for lack of a bridge over Firehole River, but the Biscuit Basin — a usage from Henderson's "Biscuit Terrace," for the knob-like formations surrounding Sapphire Pool — was hardly unknown; its thermal features had been mapped.

## MARSHALL

The surname of pioneer hotel man George W. Marshall (1846–1917) identified the stage station and the later, more adequate hotel known as the "Marshall House" at the mouth of Nez Perce Creek. From its establishment in 1880, Marshall's hospice was all there was of civilization in the geyser basins prior to 1886. But abandonment of the site after 1892 has turned that once-important settlement into little more than a vague memory.

George Marshall came to Montana Territory in gold rush days, but soon found prospecting less profitable than freighting. In 1879 he contracted to carry mail from Virginia City to Mammoth Hot Springs on the short-lived route by way of Henrys Lake. It was an unprofitable year for George, and would have been more so had he not recognized a good site for a hotel in the Lower Geyser Basin.

The hotel on the west side of Firehole River, nearly opposite the mouth of Nez Perce Creek, became the Park terminus of a stage line operated from Virginia City as the "Marshall and Goff Stage Company." The first passenger was Mrs. Carrie Adell Strahorn, wife of a guidebook publisher, who arrived in the Lower Geyser Basin on October 2, 1880, after two hard days of staging to find "the new and unfinished little log house built by Mr. Marshall, — the first and only semblance of a house in the park."

Marshall House. George W. Marshall's second hotel in the Lower Geyser Basin (in use from 1884 until 1912). Copied from "Yellowstone Park Illustrated," *The Daily Graphic*, August 11, 1888.

Of course, it wasn't the first house; there were McCartney's hotel at Mammoth Hot Springs and several other cabins in the Park, of which Mrs. Strahorn seems unaware. She continues:

> Next morning there was an early review of our surroundings; the log house was far from being finished, and the part we occupied was partitioned off with a canvas wagon cover. The second floor was only partly laid, and a window or two was missing in the upper part while the unfilled chinks between the logs allowed the rigorous October breezes to fan us at will. At that time the office and sitting room and dining room were one, and a single stove did its best toward heating the whole house.[15]

Secretary of the Interior Carl Schurz visited Yellowstone that same fall, and his arrival at the Lower Geyser Basin in bad weather was a stroke of good luck for Marshall. According to the latter's reminiscence, Schurz

> had to sleep out under a tree near my cabin one night [when] it rained. He told me next morning, 'I would have given twenty

> dollars ($20.00) to have got into a house last night,' and suggested that I should prepare to keep travelers. Said that he would see that I got a permit from the Government and when they got their leases fixed would see that I got a lease. I remained on a permit till last year when Sec. of Interior granted me a lease for 10 yrs. My first year here I did not make anything, second year came out $150.00 in debt.[16]

It was there, in that isolated cabin, that Rose Park Marshall, Yellowstone Park's first white child, was born January 31, 1881. Marshall claimed that it was Governor John Hoyt of Wyoming Territory who named his daughter: "'Rose' because roses were scarce in the Park and 'Park' because she was born in the same."[17]

After receiving his lease in 1884, Marshall built a better hotel — the "Marshall House" — on the other side of Firehole River, in the angle between the river and Nez Perce Creek. But the competition that developed after 1886 forced him to sell out to the Cottage Hotel people who in turn sold to the Yellowstone Park Association in 1889. With the completion of that company's Fountain Hotel in 1892, the settlement at the mouth of Nez Perce Creek was abandoned.

The Marshall House was the scene of a tragedy in its last years. Ellery C. Culver stayed there with his wife Mattie and little daughter, Theda, to care for the buildings during the winter of 1888–1889. Mattie was tubercular, and it was thought the mountain air would benefit her; but she worsened and died March 2, 1889, while isolated from medical help by the deep snow. There is a tradition that the men at the nearby soldier station gave what help they could (the ground was too deeply frozen for burial at that time, so they brought over two barrels that were set end to end and covered with snow to hold Mrs. Culver's body until a grave could be dug). R. R. Cummings, a maintenance supervisor for the hotel company, arrived soon after on skis on his rounds of the Park hotels, and he later described the burial:

> Finding the sick wife dead, we were compelled to remove a partition in order to secure boards with which to make a coffin. The grave was hewn down into the frozen lava formation [a gravelly soil which may have seemed like rock with the frost in it], and after two days hard work, she was consigned to rest, far away from home and friends. The wonderful character of this strong brave man, his Christian fortitude and fidelity, enabled him to

> remain and endure. . . . Alone, caring for his little baby girl, [he] remained at Fire Hole until late in April, when teams were able to reach and bring them back to their friends.[18]

Culver's daughter Theda, who was twenty months old at the time of her mother's death, was later sent to relatives in Spokane, Washington, for schooling, where she died July 21, 1906, and was buried in Fairmount Cemetery. Visitors stopping at the picnic area at the mouth of Nez Perce Creek may notice nearby a small marble headstone in a neatly fenced plot. Below a single rose, carved in low relief, are the words: "Mattie S. wife of E. C. Culver died March 2, 1889, aged 30 years." That grave and some scattered debris in the borrow pit near the road are all that remain of Marshall's brave attempt to serve Yellowstone's visitors.

## YANCEY

Two names on the Park map — Yanceys and Yancey Creek — remember a Yellowstone hotel man who probably was the area's most unusual character. "Uncle" John Yancey was an ex-Confederate who left his native Kentucky following the Civil War in such an unreconstructed mood that he "would never wear blue jeans because they reminded him of the Union Army."[19] He was a goat-bearded, shrewd-eyed, lank, Uncle Sam type (Owen Wister's description) who gave up mining on Crevice Mountain in 1882 to operate a stop-over station on the Cooke City mail route.

Upon receiving a lease in March 1884 allowing him the use of ten acres in Pleasant Valley, Uncle John began construction of his "Pleasant Valley Hotel," which was at once a wayside inn for miners and freighters, a cattle ranch, and a fishing resort. It was a primitive establishment, which Captain George Anderson thought was "not kept at a standard which would attract guests."[20] Yet he did have the patronage of a number of visitors, among them such notables as Owen Wister, Ernest Thompson Seton, and Senators Boies Penrose and George Graham Vest. They were repeatedly attracted by the renowned fishing at "Yancey's Hole" on the Yellowstone River. There is a tendency nowadays to confuse Yanceys, the hotel site in Pleasant Valley, with "Yancey's Hole," which is 1.8 miles to the north at the mouth of Yancey Creek, but they are not the same.

Yancey's hotel was said to accommodate twenty guests, but how was that done with only five rooms? The charge was $2 per day, with meals, or $10 per week. A guest of 1901 has described a stay:

Yancey's Saloon. This log saloon survived the fire that destroyed "Uncle" John's Pleasant Valley Hotel on April 16, 1906. It was razed in the mid-1960s.

We asked to be shown to our rooms. A pink cheeked little maid [daughter of the housekeeper-cook] leads the way up a stairway of creaking, rough boards and when we reach the top announces that the lady and her husband, meaning me and my daughter, can take Room No. 1. The little hallway in which we are standing is formed by undressed boards and the doors leading from it have large numbers marked upon them in chalk from one to five. An inspection of the bedrooms prove them to be large enough for a single bedstead with a box on which are washbowl, pitcher and part of a crash towel. Of the four window lights at least one was broken in every room. The cracks in the wall are pasted up with strips of newspaper. No. 1, being the bridal chamber, was distinguished from the others by a four by six looking glass. The beds showed they were changed at least twice, once in the spring and once in the fall of the year. A little bribe on the side and a promise to keep the act of criminality a secret from Uncle John induces the maid to provide us with clean sheets.[21]

The food matched the lodgings; a monotonous succession of meals featuring bacon and eggs, or fish fried in rancid butter when someone had been lucky at the river. Uncle John believed in the efficacy of slices of raw Bermuda onion which he urged on his guests at breakfast as "good for the narvs." It was the frontier plainness of his "grub pile" that brought the most complaints, causing Major John Pitcher to write him as follows:

> I am aware of the fact that it is not an easy matter for you to run a first class restaurant or table at your hotel, on account of your isolated position and distance from the railroad, but you certainly can improve on your arrangements of last summer. You have an attractive place, and in your own interests I should suggest that you make some changes and improvements in the fare which you offer travellers and tourists."[22]

A small log saloon building nearby served as the lounge for the hotel. Its plain furniture consisted of a stove, a chair, a pair of up-ended barrels supporting boards that formed a bar, and that in turn held a whiskey keg, a tin measure, and three glasses described as unwashed "since the day they had left the shelf of the crockery store." Mr. Schmide, whose comment that was, added that the whiskey was good, and Owen Wister suspected Uncle John made it so by adding "some sort of fermented stuff made from oranges he obtained from California." The saloon was also the place where Yancey slept — on the floor of the loft with his two hound dogs, according to local gossip.

Uncle John Yancey, proprietor of a truly frontier hotel, died May 7, 1903. He had gone to Gardiner, Montana, to be present on April 24 at the laying of the cornerstone for the basaltic arch then under construction. He admitted to a "warm spot in his heart" for President Theodore Roosevelt, and had come in to hear him speak. But Yancey caught cold there, which became pneumonia and carried him away. He was buried in the "Tinker's Cemetery" among other old-timers of the area.

## UNCLE TOM

While not truly a Park concessioner, Uncle Tom Richardson *was* a noteworthy institution. His guiding pleased a generation of visitors and left a mark on the map.

A visitor of the summer of 1900 mentioned that an

"Uncle" Tom Richardson. Serving a trailside lunch to his guests after guiding them to the foot of the Lower Falls of Yellowstone River by his "trail."

innovation since my last visit [1897] is the path down the cañon, which has been constructed by Mr. Richardson at the Wylie camp. He takes parties across the river in a boat, shows them the cañon from the other side [and] with the aid of rope and ladders, takes them down to the foot of the lower falls in perfect safety."[23]

H. F. Richardson, whose familiar appellation was "Uncle Tom," was a Bozeman man whose Yellowstone years began as an employee of the "Wylie Way" camping outfit. In 1896 he received his special permit to guide visitors to the foot of the Lower Falls by a route of his own devising. He rowed his dudes across the river a short distance above the rapids at the Chittenden Bridge site (there was no bridge in his day); then he took them afoot through woods yet unspoiled by the later Canyon Lodge development to a small ravine immediately east of the Lower Falls. There he had managed a "trail" of sorts, and by that means he got his parties to the bottom of the falls where they could stand close to the thundering, mist-shrouded waters with their ever-present rainbow. Uncle Tom added a good dinner — served in a sylvan setting — before rowing his charges back to the road and their transportation.

Uncle Tom did quite well with his little business. A summer usually netted him at least a thousand dollars, and that — with what he earned from Wylie before and after the season — should have made him well off; but he liked to play cards and came out of each winter broke.

The building of Chittenden Bridge in 1903, with its easy access to the east side of the Yellowstone River, ended his business. The permit under which he operated was revoked and hotel porters took over the guiding, as at other places in the Park. Uncle Tom died at Bozeman after several years of futile attempts to regain the business he had pioneered. But he is remembered in the name of the modern Uncle Toms Trail — much of it an iron stairway securely bolted to the cliff — by which visitors can go nearly to the bottom of the Lower Falls.

Now that we have had two "uncles" in a row — Uncle John Yancey and Uncle Tom Richardson — it is proper to explain that the term as used here had nothing to do with relationship; instead, it was honorific, designating a person who had been around a long time and was a likeable old bachelor. It was an endearment given only to certain older men; for instance, Yankee Jim was never so called, but Joe Brown, across the Yellowstone River from him, was always Uncle Joe Brown.

## LARRY

Unlike most place names, the one generated by this nickname was a migrant, following a particularly colorful hotel man about the Park, and it is in a sense a measure of his personal magnetism and popularity.

The places that became "Larry's" under the management of Lawrence Fancis Matthews were Trout Creek Lunch Station, on the old Mary Mountain stage road through Hayden Valley (1888–1891); West Thumb Lunch Station, which replaced the establishment on Trout Creek after the new road was built connecting the Upper Geyser Basin with Yellowstone Lake (1892); Norris Lunch Station, an ephemeral nooning — sometimes under canvas and sometimes in temporary buildings — (1893–1898), and two hotels in the Upper

Geyser Basin (1899–1904). This last included Old Faithful Inn in its first season of operation.[24] Wherever Larry went, that place very soon became indisputably identified with him.

Larry was undoubtedly the most colorful of the Park's many hosts. He was a *usually* (the qualification is necessary) genial Irishman described as "one of the fortunates who had licked the Blarney Stone." But when he wasn't genial, he was like the little girl in the nursery rhyme — "horrid"!

His glib tongue was the working end of a ready wit, and he was never at a loss for words to delight or astound his guests. Each passenger alighting from a stagecoach was met with such a salutation as, "Good morning, Governor," or "Welcome, Judge." For the Englishman it might be, "Aha, me hearty; how is the Queen?" According to Rudyard Kipling, who stopped at the Trout Creek Lunch Station in 1891,

> Larry enveloped us all in the golden glow of his speech ere we had descended and the tent with the rude trestle table became a palace, the rough fare delicacies of Delmonico's, and we, the abashed recipients of Larry's imperial bounty. It was only later that I discovered that I had paid eight shillings for tinned beef, biscuits, and beer, but on the other hand, Larry had said, 'Will I go out and kill a buffalo'? And I felt that for me, and for me alone, he would have done it. Everybody else felt that way. Good luck go with Larry.[25]

Generally, Larry pleased those who lunched with him: "Dinner was a regular circus"; "A good, plain, wholesome lunch with goodwill back of it"; and, "It is real fun to get away from the conventional big hotel." But there was another and darker side to Larry Matthews; his many years of service to Yellowstone's carriage trade had brought out something of the snob in his nature, and he had no use for those independent wagon parties — "sagebrushers" — who did not patronize the hotels, but camped by the wayside. This came out rather nastily in his altercation with the Batterman party.

Mr. C. S. Batterman complained to the superintendent that his party was abused by Larry on the afternoon of August 2, 1893, when they stopped at the Norris Lunch Station to buy some cool milk. It appears that the proprietor sized up the three couples as just another load of sagebrushers, and so did not bother to be pleasant when the men entered. They asked for milk and were told there was none for sale, but that they could buy beer. Declining that, Mr. Batterman carried some water to the ladies waiting in the wagon.

When the water glasses were returned, Larry did produce a pitcher of milk that he offered at twenty-five cents a glass. But Mr. Batterman refused such an extortionate price, whereupon Larry proceeded to charge him twenty-five cents for each glass of water. Along with the tab, which was also refused, came "a torrent of vile language" and an order to get out. The men were followed to their wagon by the angry Irishman, who "informed us and the ladies that no pricks were allowed in his house, but that it was exclusively for gentlemen and that if we dared set foot in his place he would send for the soldiers and have us driven out of the park."[26] Though Mr. Batterman was not a gentleman in Larry's eyes, he did get the attention of several congressmen.

Toward the end of his eighteen years in the Park, in which he moved up from proprietor of canvas-covered lunch stations to hotel management, Larry became overly conscious of his importance and less often polite and courteous; also, he was more likely to yell at such people as Joe, the yardman, which explains this bit of doggerel by a former employee:

Oh yes, they say it's tru enuf —
Old Faithful's quit.
'Twas on account of Larry's guff,
The Divel couldn't stand it.
He was poking in a Chicago batch
Just to make the old thing go,
When far off sounds his ear did catch;
'Twas Larry calling Joe.
He dropped his shovel, the fire went out
— and Faithful quit![27]

That was Larry, who for years was as much of an attraction as Old Faithful Geyser. He retired after the 1904 season, leaving — as Campbell's guidebook later said — "a fragrant memory of his jibes and jokes."

## MISCELLANY

As mentioned earlier, George L. Henderson, the early-day Park policeman who became a hotelman, is the source of many Park place names.

However, it is unclear whether he originated the names or merely preserved current usages by publication in his newspaper-like guide to Park features.[28] Unfortunately, Henderson's descriptions are often too vague to allow a sure correlation of his names with known features.

In addition to the two place names already discussed that had origins in Henderson usages — Black Growler Steam Vent and Sapphire Pool — there are fourteen others that stand as the essential sources of approved usages. These surviving place names, a mere 7.7 percent of the total of 209 that appear to have originated with Henderson, include such important ones as Hymen, Jupiter, Narrow Gauge, and Minerva Terraces at Mammoth Hot Springs; the Hurricane Vent and Porcelain Basin at Norris; the Jewel Geyser and Silver Globe Spring in Biscuit Basin (his "Biscuit Terrace"); and Golden Gate, Kingman Pass, and Elk Park along the road to the geysers.

Fifty-eight of Henderson's names (27.7 percent) are no longer in use — though the features to which they were applied are known — most going by prior or more appropriate designations. In a few cases, changing conditions have rendered naming quite unnecessary. "Art Spring" disappeared when "Cupid's Cave" was blocked by hot spring deposits. And there was no further interest in the "River Styx," underneath the Devils Kitchen, after that solution hole in the upper terrace formation was closed to visitors because of the danger from irrespirable gases. "Clark Town," "Haynes Town," and "Henderson Town" at Mammoth Hot Springs are other names that disappeared with the concession enclaves they once designated. "Lamartine Road" and "Swanson Road," once used to distinguish the sections of roadway built by the army engineers below and above Mammoth Hot Springs, had no meaning for later generations without memory of those mighty foremen (actually, they were both crooks, but that wasn't apparent for a time). Other Henderson road names that became obsolete are "Belt Line" for the traffic loop circling the Park, now the Grand Loop Road (1923); "Seven Lake Avenue," the new road connecting Obsidian Cliff with Norris Geyser Basin in place of the section of the Norris Road by way of Lake of the Woods; "National Park Avenue," the new road through the Lower Geyser Basin, renamed "Fountain Freight Road" (1930) and now only a scenic access; and "Split Rock Bridge," made unnecessary by the construction of a road under the cliffs on the east side of Gardner River.

Many of the features Henderson named already *had* names, as was the case with his "Boiling River," where the heated runoff from

Mammoth Hot Springs discharged into Gardner River. The Geological Survey had called that steaming flood Hot River the previous year (1884), and so it remains. Nor did "Canyon Avenue" take for the new road from Norris Geyser Basin to the falls of Yellowstone River; it had already (1884) appeared in the public press as Norris Cutoff. Henderson's "Falls of the East Gardiner" had been named Undine Falls by the Jones party twelve years earlier, and his "Falls of the West Gardiner" was only the Rustic Falls named by Superintendent Norris six years earlier (1879). Even in the matter of the "Falls of the Middle Gardiner," Henderson did not prevail over the Geological Survey's decision to use Osprey Falls the same year (1885). Similar names were Henderson's "Hayden Park," the Hayden Valley named by Lord Dunraven (1874); his "Limestone Mountain" had been Terrace Mountain before 1885; his "Monument Mountain" was the same that Captain Barlow had called Sepulcher Mountain (1871) — the name it still bears; "The Pyramids" had been named Pyramid Peak by the Geological Survey (1883); and his "White Elephant" — present White Elephant Back (1922) — was commonly known as "White Elephant Geyser" two years earlier (1883).

Other Henderson names were not appropriate and were replaced with better usages. His "Capitol Hill Terrace" became a less cumbersome Opal Terrace (1936); the "Colored Paint Pots" on the edge of the Gibbon Geyser Basin became Artists Paintpots (1930); and the "Great Paint Pots," which was a cut better than the plain "Mud Pots" of the Hayden Survey (1871) or the equally plain "Paint Pot" of Col. Gregory (1881)[29], at last became Fountain Paint Pot (1927). But there were several other tries at naming before the superintendent's Place Name Committee came up with that one. Wingate had called the forty-by-sixty-foot basin of bubbling, plashing mud a "Mud Caldron" (1886), noting that "it does not boil violently, but simmers peacefully, while all over its surface bubbles of steam constantly form and break with a curious *blob blob*." He also mentioned the "small craters or mud pots of the most delicate shades of pink and rose."[30] Park photographer Frank J. Haynes used "Mammoth Paint Pots" (1889) in his guidebook and on the stereographs he sold, and that name almost made the grade. But the obvious confusion with the Mammoth Hot Springs name led to a resolution based on the word "Fountain" (from the nearby thermal area).

Henderson also tried to use the word "Mammoth" in naming Lower Geyser Basin features. His "Mammoth Geyser," for the Great

Fountain Geyser (so called by the Hayden Survey in 1872), was a dud for the same reason as the "Mammoth Paint Pots" name, and so too, his "Mammoth Geyser Basin" — a partitioning of the Lower Geyser Basin that was quite unnecessary. Personally, I dislike the practice of recognizing basins within basins, as in the Upper Geyser Basin, where there are two subareas called Black Sand Basin and Biscuit Basin (we can blame Henderson for the latter, as it is only an alteration of his "Biscuit Terrace").

At Mammoth Hot Springs, Henderson had a real problem with his terrace names. He also referred to the present Minerva Terrace (his name) as "Cleopatra Terrace," "Diana Terraces," and "Coating Terrace." The latter usage referred to the coating of souvenir items — horseshoes, rings, baskets, toys, and other things — with alabaster-like lime deposit by exposing them on racks over which the thermal waters were directed. Visitors could leave their trinkets at Henderson's store to be coated while they were in the Park, or they could buy coated objects kept for sale there (Ole Anderson continued that activity after he took over the curio business). One of Henderson's terrace names — "Cedar Terrace" — was a complete misnomer, for the area it referred to was not a terrace but a belt of trees on a steep slope between two old hot spring terraces (the distinct sections of the present Mammoth Hot Springs settlement); perhaps that is why the name disappeared.

Several of Henderson's usages are puzzling. Why did he call Black Warrior Spring "The Marguerite" in 1885, and then change it to "Black Warrior Geyser" in 1891? For whom or for what was his "Walpurgia Lake" named in 1885? Was it a literary allusion to the nightmarish wildness once associated with the eve of May Day, the *Waldpurgisnacht* of Medieval Europe? (Perhaps the answer is moot, because the Hot Lake name given that feature by the Hayden Survey seven years earlier has remained the approved usage.) Similarly, his use of "Welton Park" for Gardners Hole, the circular valley on the headwaters of Gardner River, is unexplained. The men of the fur trade knew that beautiful valley as the favorite haunt of Johnson Gardner (see Chapter 2), but it did not bear his name on Park maps until 1958.

Almost two-thirds of Henderson's place names (64.6 percent) were given to features that cannot now be identified within a reasonable degree of certainty; therefore, they will be discussed in Chapter 10 with "Those That Didn't Make It."

The publisher of another Park guide, photographer Frank Jay Haynes, made a considerable contribution to Yellowstone's place names. As in Henderson's case, unfortunately, it is sometimes impossible to tell if he was the originator of a name or only the first to put it into print. Of the sixty-one usages that appear to be his, fifteen (24.6 percent) are approved, with only one of those at Mammoth Hot Springs: Angel Terrace (1896). This is a white deposit cascading down through standing dead trees near the start of the loop road around the upper terraces. It must be added that G. L. Henderson did his utmost to confuse the naming at this point. In 1883 he used the name "Bethesda Plateau" for the whole formation, on the presumption that its waters had healing powers similar to those of the Biblical pool (John 5:2–4). The same year he named part of it "Chameleon Terrace" for the observed color changes, while another part was named "Rath Spring and Terrace" for the German geologist G. M. von Rath, a visitor of that year. Other names Henderson applied to portions of this formation were "Ingersoll Spring and Terrace" (1884) for Robert G. Ingersoll, politician and orator; "Haynes Terrace" and "Minnehaha Terrace" (1885); and finally, "White Sulphur Spring Plateau" (1885), again for the entirety.

Returning to the names for which F. J. Haynes appears to be responsible, he did no better in the Norris Geyser Basin — just one name, Iris Spring (1915), can be traced to him; and at the Lower and Midway basins, none. He did name some other features in the Lower Geyser Basin that cannot now be located with certainty. "Buffalo Spring," near the Great Fountain Geyser, was likely the same spring in which David E. Folsom found "the entire skeleton of a buffalo that has probably fallen in accidentally and been boiled down to soup."[31] Neither "Firehole Spring" nor "Mushroom Spring" is described, though the present Firehole Lake would seem a good possibility for one. "Silex Pool" is just a play on the well-established name Silex Spring, first used by the Geological Survey (1904).

In the Upper Geyser Basin, more of F. J.'s names remain in use. There is Artemisia Geyser (his "Artemisia Spring," and the same feature the Hayden Survey called "Restless Geyser" in 1878); Cascade Geyser; Economic Geyser Crater (his "Economic Geyser"); Emerald Pool; Mortar Geyser; Pump Geyser; and Splutter Spring (his "Splutterer Spring"). Also, as noted in Chapter 5, F. J. was instrumental in popularizing the name of Lone Star Geyser, although he did not originate it.

A few other approved names are scattered about the Park. These are: Gibbon Canyon, Grand View, the Petrified Tree, at the head of Yancey Creek (this was originally "Petrified Trees," but one of the two stumps was not fenced and has been entirely removed by souvenir pickers), and, of course, Yanceys — the hotel site, not the fishing hole on Yellowstone River. This was merely a contraction of the name Uncle John gave his establishment in 1884 ("Yancey's Pleasant Valley Hotel").

Frank J. Haynes, Yellowstone's official photographer from 1883 until he turned his Park business over to his son Jack E. Haynes in 1916, died March 10, 1921. Superintendent Horace M. Albright immediately designated an eminence on the south wall of Madison Canyon (the bluff unofficially known as Mount Burley after 1912, for D. E. Burley of the Union Pacific Railroad) as Mount Haynes. In 1935, Superintendent Roger W. Toll placed the name of William Henry Jackson, the man who made the first photograph in Yellowstone Park (1871), on a height on the north wall of the canyon, nearly opposite Mount Haynes. Jackson, whose life work centered on pictorial documentation of the post–Civil War West, lived seven more years, dying in 1942 at the advanced age of 99 years.

Thus far the record has lacked even a hint that the minor employees of the Park concessioners named things — which they most certainly did. So here, in the "Merry Widow Geyser," is one given to a very unusual thermal feature near Old Faithful Inn, by the winter keeper (name unknown) during the winter of 1911.

> Only lately a hitherto inactive hot pool broke into sudden activity. Above it had been reared a tent. Its surface was covered with a floor through a trapdoor in which its hot water was raised into washtubs. It was surmounted by the laundry of Old Faithful Inn. During the winter when none was there to witness the eruption except the winter keeper, the explosion came. He was entering his winter greenhouse nearby when, with a sudden roar, the hiss of steam, and the trembling of the earth, the laundry and all its contents, floor, tubs, boxes, and benches, were tossed skyward. . . . The pool had become a geyser, and with a thought of popular celebrity the single witness promptly named it the Merry Widow.[32]

However, the "Merry Widow Geyser" turned out to be a one-time performer; once rid of the laundry, it returned to being just a slumbering hot pool.

## NOTES

1. Land Records of Gallatin County, Horr, H. R., 1871, July 5, Decl., Mammoth Hot Springs. H–309. Gallatin County Courthouse, Bozeman, Montana.
2. That name is essentially the same as one used in the first reference to this feature, "White Sulphur Banks," which appeared on Captain Washington Hood's map (1839). According to Lt. J. W. Gunnison (1852), the trappers knew them as "The Great Springs"; and Capt. W. F. Raynolds (1860) showed a "Sulphur Mt." at just the right place on his map. The Barlow-Heap party of 1871 used "Soda Mt.," and the Hayden Survey parties of that year couldn't make up their minds as to whether the name should be "White Hot Springs" or "White Mountain Hot Springs."
3. "Hot Springs . . . ," Bozeman, (MT) *Avant Courier*, Sept. 13, 1871.
4. Letter, H. R. Horr to Capt. Geo. L. Anderson, May 16, 1891. Yellowstone Archives, Doc. 1050. Yellowstone Park Reference Library.
5. Hiram M. Chittenden, *The Yellowstone National Park*, 4th ed. (Cincinnati: Robert Clarke Co., 1903), p. 280.
6. W. H. Dudley, *The National Park from the Hurricane Deck of a Cayuse, or The Leiderkrantz Expedition to Geyserland* (Butte City, MT: Free Press Publishing Co., 1886), p. 41.
7. C. Hanford Henderson, "Though the Yellowstone on Foot," in *Outing* (May 1899), p. 161.
8. "Local Layout," Livingston (MT) *Enterprise*, August 1, 1885.
9. "Park Items," Livingston (MT) *Enterprise*, November 21, 1885.
10. George L. Henderson, *Yellowstone Park Manual and Guide* (published in newspaper format, July 1, 1885), p. 1.
11. A. B. Guptill, Ed., *Haynes' Guide to Yellowstone Park* (St. Paul, MN: F. J. Haynes, 1896), p. 34.
12. Jack Ellis Haynes, *Haynes Guide: Handbook of Yellowstone National Park* (Bozeman, MT: Haynes Studios, Inc., 1966), p. 66.
13. Ferdinand V. Hayden, *Sixth Annual Report of the United States Geological Survey of the Territories . . . for the Year 1872* (Washington: Government Printing Office, 1873), p. 672.
14. George L. Henderson, *Yellowstone Park: Past, Present, and Future* (Washington: Gibson Bros., Printers and Bookbinders, 1891), p. 9.
15. Carrie Adell Strahorn, *Fifteen Thousand Miles by Stage*, 2nd ed. (New York: G. P. Putnam's Sons, 1915), pp. 258–259.
16. G. W. Marshall, Biographical statement in the Bancroft Library [probably written in 1885]. From a photostat furnished Yellowstone National Park October 12, 1938.
17. Ibid.
18. News item, "Passing of Ellery C. Culver," from an unidentified newspaper (1922); copy with family papers provided the author January 11, 1990, by Nan Weber, Salt Lake City, Utah.
19. Owen Wister, "Old Yellowstone Days," *Harpers Monthly*, 172 (March 1936), p. 479.
20. George Anderson, *Report of the Acting Superintendent of the Yellowstone National Park to the Secretary of the Interior, 1896* (Washington: Government Printing Office, 1896), p. 9.

21. Carl E. Schmide, *A Western Trip* (Detroit, MI: Herold Press, 1910), pp. 59–60.
22. Letter, John Pitcher to John Yancey, May 31, 1902. Letters Sent, Vol. 11, p. 396, Yellowstone Archives, Doc. 681.
23. Henry T. Finck, "Yellowstone Park as a Summer Resort," *The Nation*, 71/1839 (Sept. 27, 1900), pp. 249–250.
24. L. F. Matthews, letter to manager J. H. Dean, October 1, 1898; Northern Pacific Railroad Files, "President, 210-A."
25. Rudyard Kipling, *From Sea to Sea* (Garden City, NY: Doubleday, Page & Co., 1920), p. 100.
26. C. S. Batterman to Captain Anderson, August 2, 1893. Yellowstone Archives, Doc. 681.
27. Bill Scales, letter to the Chief Ranger, Yellowstone National Park, May 20, 1963, p. 6.
28. *Yellowstone Park Manual and Guide* (privately printed, July 1, 1885), 4 pp. This 1st edition sold for 25 cents, while a two-page 2nd edition, in 1888, was priced at 10 cents.
29. James F. Gregory, "Journal," dated October 20, 1881; printed as pp. 12–23 with the *Report of Lieut. General P. H. Sheridan*, Sept. 20, 1881 (Washington: Government Printing Office, 1882), p. 20. Gregory refers to the present Fountain Paint Pots as a "curious boiling clay-geyser called the Paint Pots."
30. George W. Wingate, *Through the Yellowstone National Park on Horseback* (New York: O. Judd & David Judd, 1886), p. 99.
31. Charles W. Cook, David E. Folsom and William Peterson, *The Valley of the Upper Yellowstone*, ed. by Aubrey L. Haines, (Norman: University of Oklahoma Press, 1965), p. 44. The entry is for October 2, 1869.
32. John H. Raftery, *The Story of Yellowstone* (Butte, MT: McKee Printing Co., 1912), pp. 123–124.

# 8

# SPECIAL NOTICE FOR VISITORS



The Yellowstone region had visitors — persons whose reason for going there was not commercial but mere curiosity — before it became a park. It is easy enough to distinguish the early visitors from those who came for business reasons (as the fur hunter or prospector); but there is some difficulty in separating them from the explorers. I prefer to think of explorers as those whose curiosity improved the general knowledge of the area as known to the public, and visitors as those who only satisfied a personal curiosity. It is the noncommercial, passive role that defines the visitor for the purposes of this chapter.

Warren Angus Ferris, a clerk of the American Fur Company, took advantage of a slack time in his fur trade activity to visit a geyser basin on Firehole River with two Pend d'Orielles Indians in 1834, and Father Francis Xavier Kuppens, a young Jesuit priest attached to the old St. Peters Mission on Sun River, entered what is now Yellowstone Park with a band of Piegan Indians in the spring of 1865. But it was only recently that these early visits were generally known.

The vanguard of that visitation that now exceeds two million persons each summer season appeared in 1871 close on the heels of the Hayden-Barlow explorers, and months ahead of the enactment of the legislation creating Yellowstone National Park on March 1, 1872. The visitors who thus anticipated establishment of a public park in "Wonderland" (they used that term in a series of newspaper articles) left Deer Lodge, MT, on August 7, 1871. The party consisted of U.S. Mining Commissioner R. W. Raymond and his assistant, a Mr. Eiler; A. F. Thrasher, the town's photographer; C. C. Clawson, a reporter for the Deer Lodge *New North-West*, and J. S. Daugherty, an Indiana businessman who was "on a tour of sight seeing and for his health." They went by way of Virginia City and Henrys Lake, where they picked up Gilman Sawtell as their guide, and they returned to Deer Lodge after a series of adventures that included loss of much of Thrasher's photographic equipment in a forest fire. They arrived home in time for Clawson to get the first installment of his serialized account — "Notes on the Way to Wonderland" — in the September 9 issue of his newspaper.

The Bozeman *Avant Courier* of February 29, 1872, mentioned the imminent departure of a party "consisting of James Pencil and lady, N. B. Ford and Julius Levi, of Helena, and H. C. Philbrook, Ike Fordonski and Johnny Kennedy, of Bozeman" for the Mammoth Hot Springs; thus, they were undoubtedly the first visitors to the new and yet nameless national park. The primitive nature of the roads in the vicinity of Yellowstone Park, combined with the even rougher routes within the area, made a Park visit an arduous business prior to construction of the Northern Pacific's Park Branch line in 1883. That worked to keep the visitation essentially local.

These local people came with their own outfits, by way of Bozeman and Yankee Jim's toll road to Mammoth Hot Springs, or by the "Virginia City and National Park Free Wagon Road" to the geyser basins on the Firehole River. It was such visitors that Captain Ludlow found prowling about the Upper Geyser Basin in 1875, "with shovel and axe, chopping and hacking and prying up great pieces of the most ornamental work they could find. Men and women alike joining in the barbarous pastime." But why should they know better? No one had yet told them they shouldn't enrich their what-not cabinets with choice specimens from the Park!

The coming of the railroad to Montana and the Park, and the concomitant improvement within the area through road building and introduction of such amenities as stage lines and hotels, rapidly changed the character of the visitation. "Tourists" — persons sightseeing on paid-in-advance tour tickets — were delivered to the Park by the railroad, transported on scheduled stagecoaches, and lodged more or less comfortably in hotels. The "carriage trade," as this way of seeing the Park was styled, was more genteel than the local visitation had been, drawing heavily from the eastern states and Europe, and it lasted a little more than three decades. During that time the "sagebrushers," local people traveling with their own outfits — the "great unwashed" in the eyes of concessioners — were often victims of discrimination.

The introduction of a new form of transportation — the private automobile — was required to democratize the Park. Many changes followed in the wake of that Ford roadster from Minnesota that received permit No. 1 at the North Entrance on July 31, 1915: campgrounds appeared throughout the Park, then cabins and tent facilities, gas stations, cafeterias, bathhouses, and wood yards. As the concessioners were forced to orient their businesses more toward the automobilists, lodge developments followed for those who wanted something between austerity and the hotels. It was obvious quite soon after World War II that the carriage trade was history in Yellowstone Park, and that the future belonged to the automobilist.

The influence of visitors on Yellowstone's place names has been slight, but not for lack of trying. A cursory review of Park literature has discovered 189 names that can be attributed to visitors, and that is undoubtedly far from all; and yet, only 29 have gained acceptance (a number of those because they served an official purpose). The impermanence of most names given by visitors is a result of some inherent defect or a lack of support, for place names survive in the long run only by official approval or common usage. Many of these maverick names were unacceptably personal, frivolous, or in bad form, and they were often applied to features that already had accepted names. The result could be confusing, particularly where such names became embedded in the literature of the Park. As an example, Crested Pool, adjacent to Castle Geyser in the Upper Geyser Basin, was described by Thomas in *The Graphic* (1888) under the name of "Diana's Well"; and by Guptill in his 1889 guidebook as "Castle Well." An undated stereographic photograph distributed by C. R. Savage was titled "Crystal Spring," and as late as 1919 Alma White used "Castle Spring" in her book, *With God in the Yellowstone*. These are all tourist monikers for a feature that had an appropriate name from 1878 on!

Park visitors of the Victorian era were particularly fond of place names with a satanic connotation (probably because they grew up with a "hellfire and brimstone theology" and were able to relate it to what they saw in the Park). As a result, thirty-nine features within the Park were given devilish names, or fifty-one if names based on "hell" are included. As Charles T. Whitmell said in 1885: "Could Dante have seen this region he might have added another terror to his Inferno." Only six names of that type remain in approved use, and they are relics of Yellowstone's opening years.

## FERRIS

This surname, which identifies a fork of Bechler River, has the distinction of being the Park's only place name given by a visitor to honor a visitor.

A direct result of exploitation of the Snake River Valley's agricultural potential during World War I was a demand for more irrigation water in the postwar years. Idahoans thought that the increase could only come from Yellowstone Park's water-rich Cascade Corner, and that view was the basis of a campaign mounted in 1920 by irrigation interests seeking to turn that part of the Park into an irrigation reservoir (justifying their proposal by pointing out that the Cascade Corner — the basin of the Bechler and Fall Rivers — was mapped as swamp and therefore of little value as recreational land).

Fortunately, Yellowstone Park had a friend in the National Parks Association, and a member of that organization, William C. Gregg of Hackensack, New Jersey, was so concerned that he devoted three summers to an investigation of the park potential of the Cascade Corner. He found that the 1884 topographic map was in error in presenting the basin below the mouth of Bechler Canyon as a swamp; in fact, he found a delightful area of meadow and forest, containing many beautiful streams and waterfalls — a most inviting place for the fisherman and back-country camper. He also found its features largely unnamed.

Gregg opposed the irrigation interests in articles published in *The Outlook* magazine in 1921 and 1923, and in those he gave names, which have mostly survived, to a number of features. He said: "Our search [for men to honor with place names] was rewarded with two — Ferris, who visited the Geyser Basins around 1830, and was the first man to write about it, and Phillips [W. Hallett], one of the defenders of the Park thirty years ago. We named two streams for them."[1]

Although the Ferris account is not the first description of a Yellowstone Geyser to appear in print (the letter of Daniel T. Potts, first published anonymously in the *Philadelphia Gazette and Daily Advertiser*, September 27, 1827, and reprinted by *Niles Register*, has that honor), it remains interesting, and worthy of space here. Ferris begins on May 18, 1834:

> I had heard in the summer of 1833, while at rendezvous, that remarkable boiling springs had been discovered, on the sources of the Madison, by a party of trappers in their spring hunt; of

which the accounts they gave were so very astonishing, that I determined to examine them myself, before recording their descriptions, though I had the united testimony of more than twenty men on the subject, who all declared they saw them, and that they really *were*, as extensive and remarkable as they had been described. Having now the opportunity of paying them a visit, and as another or a better might not soon occur, I parted with the company after supper, and, taking with me two Pend'orielles, set out at a round pace, the night being clear and comfortable. We proceeded over the plain about twenty miles, and halted until daylight on a fine spring, flowing into Camas Creek. Refreshed by a few hours sleep, we started again after a hasty breakfast, and entered a very extensive forest called the Piney Woods [Madison Plateau], which we passed through, and reached the vicinity of the springs about dark, having seen several small lakes or ponds [Summit and Little Summit lakes], on the sources of the Madison; and rode about forty miles; which was a hard day's ride, taking into consideration the rough irregularity of the country through which we had travelled.

We regaled ourselves with a cup of coffee, and immediately after supper lay down to rest, sleepy, and much fatigued. The continual roaring of the springs, however, for some time prevented my going to sleep, and excited an impatient curiosity to examine them; which I was obliged to defer the gratification of, until morning; and filled my slumbers with visions of water spouts, cataracts, fountains, *jets d'eau* of immense dimensions, etc. etc.

When I arose in the morning [May 20, 1834], clouds of vapor seemed like a dense fog to overhang the springs, from which frequent reports or explosions of different loudness, constantly assailed our ears. I immediately proceeded to inspect them, and might have exclaimed with the Queen of Sheba, when their full reality of dimension and novelty burst upon my view, 'The half was not told me.'

From the surface of a rocky plain or table, burst forth columns of water, of various dimensions, projected high in the air, accompanied by loud explosions, and sulphurous vapors, which were highly disagreeable to the smell. The rock from which these springs burst forth, was calcareous, and probably extends some distance from them, beneath the soil. The largest of these

> wonderful fountains, projects a column of boiling water several feet in diameter, to the height of more than one hundred and fifty feet, in my opinion; but the party of Alvarez, who discovered it, persists in declaring that it could not be less than four times that distance in height — accompanied with a tremendous noise. These explosions and discharges occur at intervals of about two hours. After having witnessed three of them, I ventured near enough to put my hand into the water of its basin, but withdrew it instantly, for the heat of the water in this immense cauldron, was altogether too great for my comfort; and the agitation of the water, the disagreeable effluvium continually exuding, and the hollow, unearthly rumbling under the rock on which I stood, so ill according with my notion of personal safety, that I retreated back precipitately, to a respectful distance. The Indians, who were with me, were quite appalled, and could not by any means be induced to approach them. They seemed astonished at my presumption, in advancing to the larger one, and when I safely returned, congratulated me on my 'narrow escape.' They believed them to be supernatural, and supposed them to be the production of the Evil Spirit. One of them remarked that the hell, of which he had heard from the whites, must be in that vicinity. The diameter of the basin into which the waters of the largest jet principally fall, and from the centre of which, through a hole in the rock of about nine or ten feet in diameter, the water spouts up as above related, maybe about thirty feet. There are many other smaller fountains, [but] they did not throw their waters so high, but occurred at shorter intervals. In some instances, the volumes were projected obliquely upwards, and fell into neighboring fountains, or on the rock or prairie. But their ascent was generally perpendicular, falling in and about their own basins or apertures. These wonders of nature are situated near the center of a small valley surrounded by pine-crowned hills through which a small fork of the Madison flows.[2]

More will be said later about Gregg's naming of Yellowstone features; it is sufficient now to add that Superintendent Albright, in 1930, put Gregg's name on the middle fork of the three that come together at Three Rivers Junction to form the Bechler River. That was a well-deserved tribute to a visitor's concern for Yellowstone Park.

## MARY

This given name may be the only one from visitor sources that has thrived without official nurturing, perhaps because the circumstances behind it were so well recorded.

Mary Clark, the young lady for whom Mary Lake on the divide between Alum and Nez Perce creeks is named, was a member of a party that entered the Park from Bozeman in August, 1873. Just who led this party is uncertain, though it certainly included talent enough. In addition to the Rev. Edwin J. Stanley, a Methodist minister who later wrote a book about the excursion, there were Granville Stuart of Deer Lodge; Col. W. F. Sanders, a Montana politician; Hiram Knowles, a Montana Judge; Professor Egbert Smith; W. T. Todd, of Helena; and a Mr. Robbins of New York. All we know about Mary Clark is that her home was in Chicago.

The new Park was yet a virtual wilderness with primitive trails requiring saddle and pack-horse travel beyond Mammoth Hot Springs, where the rude cabins of McGuirk and Horr and McCartney represented the ragged edge of civilization. The route on in to the geyser basins was by the old Indian trail to Tower Fall, then over the Washburn Range by way of Dunraven Pass into Hayden Valley, and from there up Alum Creek to a crossing of the Central Plateau and

Mary Lake Patrol Cabin. This typical back-country patrol cabin faces the lake named for visitor Mary Clark on August 29, 1873. Both cabin and lake are on Mary Mountain.

down Nez Perce Creek to the Lower Geyser Basin. That was the standard route, although Woodruff and Topping had found a shorter one by way of the Norris Geyser Basin the year before (1872) and had used it to guide a party to the geyser basins that included Mrs. Hiram H. Stone — the first white woman to get that far.

The Rev. Stanley wrote of the crossing of Central Plateau on August 29, 1873: "We passed along the bank of a lovely little lakelet. . . . It is near the divide, and on its pebbly shore some members of our party unfurled the Stars and Stripes, and christened it Mary's Lake, in honor of Miss Clark, a young lady belonging to our party."[3] The name has been standardized as Mary Lake in accordance with the rule that drops the possessive on place names.

As clear as that record is concerning the origin of this name and its usage, there have been some problems. Henderson got it wrong when he stated, "Mary's Lake and Mountain were named and consecrated by Superintendent P. W. Norris to a beloved school mate whose memory he cherished."[4] Of course, that is just romanticized malarkey.

The height of land upon which Mary Lake lies has been officially known as Mary Mountain since 1895 (it does not appear to be a mountain as approached from the east, but the west side drops abruptly into Nez Perce Creek, leaving no doubt that the name is appropriate). General Howard's military road, which was hastily built over Mary Mountain during the Nez Perce War (1877), was used as a stagecoach route until 1892, but the building of a road from Old Faithful to Lake by way of West Thumb made the difficult "Mary Mountain Road" obsolete. It was abandoned for vehicular use after World War II and is now the Mary Mountain Trail.

## ELEANOR

This given name, used at least as early as 1914 to designate a small lake almost in the throat of Sylvan Pass, is one of Yellowstone's inscrutable mysteries.

Eleanor Lake — often called "Eleanor Pool" because it serves as a sump for meltwater cascading down from snowbanks high on Top Notch Peak — made its initial appearance in the 1914 edition of *Campbell's New Revised Complete Guide . . . to Yellowstone Park*, p. 132 (a photograph of "Eleanor Cascade," the seasonal water source for the lake, is on page 99); however, the guidebook does not indicate the origin of the name. I have examined all the reports of the United States Engineer Department from the initiation of the East Entrance Road

Eleanor Lake. Eleanor's identity is uncertain, but she might have been the daughter of Captain Hiram M. Chittenden, the engineer officer who directed work on the Sylvan Pass road.

project to its completion without finding the lake referred to by name. It is tempting to think that perhaps Captain Chittenden gallantly did as was done at Craig Pass — used the name of some lady who passed by at the right time — but then, he had a daughter by that name. Was it she, Eleanor Chittenden Cress, rather than a visitor? We will likely never know.

## HAYDEN/DUNRAVEN

These surnames, which appear in names of two outstanding Park features, are strangely related: the one was given *by* a visitor and the other *for* that same visitor.

Hayden Valley, the beautiful meadowland through which the Yellowstone River meanders between the outlet of its great lake and the two falls, is the exposed bottom of a primordial lake of greater extent than present Yellowstone Lake. Evidence of that ancient lake, which undoubtedly discharged its water into the Snake River drainage by way of a low pass at the head of Outlet Creek, is seen in the old beach terraces so prominent at Terrace Point.

This magnificent feature immortalizes Ferdinand Vandiveer Hayden (1829–1887), a medical doctor turned geologist who organized the United States Geological Survey of the Territories and headed it for a decade — until the several independent federal surveys engaged in mapping the West were merged in 1879 to form the present United States Geological Survey. Dr. Hayden was an outstanding geologist whose fault, if indeed it was a fault, lay in attempting to do too much too fast. Historian Chittenden has this to say about him:

> His desire to forward the cause of science was sincere and enthusiastic, and he was always ready to modify his views upon the presentation of evidence. He was intensely nervous, frequently impulsive, but ever generous, and his honesty and integrity undoubted. The greater part of his work for the government and for science was a labor of love."[5]

Exactly who was responsible for the naming remains uncertain. According to Chittenden, it was the "U.S.G.S." in 1878, but that organization did not come into existence until a year later. An examination of the Hayden Survey's Park-wide map of 1878 — in a manner of speaking, the final accounting of their cartographic effort in the Park — shows no name on Hayden Valley, and the map accompanying Superintendent Norris's *Report* for the year 1878 is also lacking in a name for what is now Hayden Valley. But there is a "Haydens Valley" on the more detailed map accompanying Norris's 1880 *Report.*

The possibility that Norris may have picked up the name from a book published by Dunraven following his American visit finds support in the reminiscence of railroad surveyor Panton: "Hayden said he had never named anything for himself, claiming Hayden Valley was named by Lord Dunraven on his tour with Texas Jack in 1872 [1874]."[6] Until a better explanation is found, that one — which is in accordance with Hayden's naturally self-effacing nature — will have to do.

Dr. Hayden's name had previously (1872) been given by Bradley to the stream then known as the "East Fork of Firehole River," but "Hayden's Fork" did not catch on. This is perhaps just as well, because events yet waiting in the wings would supply a better name by identifying that stream as Nez Perce Creek — for the Indians who passed that way on their valiant retreat across the Park in 1877. They were visitors also.

The names honoring Lord Dunraven — Dunraven Peak, given by Henry Gannett (1878), and Dunraven Pass (1879) — have been covered in Chapter 5 (see Dunraven).

## ROOSEVELT

There are a number of good reasons why the man who was undoubtedly the Park's most famous visitor, President Theodore Roosevelt, should be honored on the map; and he was — doubly — though there is a bit of deception in one usage.

"Teddy" Roosevelt, our rough-rider president, visited the Park in the spring of 1903, just as the foundations for the great stone arch intended as a "suitable entrance gate" for tourists arriving by rail at the North Entrance were being laid. With the work at the right stage for placing the cornerstone, Charles W. Miller, the hotel company's manager, approached Judge Meldrum with the idea of asking the President to dedicate the arch under Masonic auspices.

Meldrum communicated that idea to Major Pitcher, who was the President's host in the Park, and received this reply on April 14:

> Dear Judge:
> The President directs me to say that it will give him great pleasure to comply with the request of your committee concerning the laying of cornerstone at the entrance gate to the Yellowstone National Park, on April 24, 1903.
>
> Captain Chittenden will see you tomorrow evening and will arrange the details for the ceremony.[7]

The dedication, which was the final event of the President's strenuous two-week "vacation," was attended by more than two thousand persons (many of them brought up from Livingston on four special trains). The ceremonies began at three in the afternoon on Gardiner's main street, with marching Masons and music by the Livingston band. The laying of the cornerstone over a recess filled with mementoes of the occasion was followed by a speech in which Roosevelt said all the things his audience wanted to hear, from the uniqueness of the great Park he had just visited to how good it was to be back in "this great western country."[8]

The imposing stone structure with the welcoming words, "For the Benefit and Enjoyment of the People" over its portal seems to have gone without a name until 1905, when Superintendent John Pitcher noted that six of the twelve small Sequoia trees shipped to the Yellowstone from Sequoia National Park "have been planted near the Roosevelt Arch," where they were expected to be of interest to future tourists.[9]

Park guidebooks referred to the arch as "North Entrance Arch" and "the arch at North Entrance" until 1942, when the *Haynes Guide* used Theodore Roosevelt Arch (pp. 35–36). The name was used that way to the end of publication of the *Guide* in 1966. The simpler usage — Roosevelt Arch — should be given preference.

Three years after President Roosevelt's visit to Yellowstone Park the Wylie Permanent Camping Company ("Wylie-Way"), by then under the management of A. W. Miles of Livingston, Montana, established a tent camp on Lost Creek, two miles northwest of Tower Falls. This camp, which provided primitive overnight accommodations for traffic using the new road opened in 1906 over the Washburn Range, went unnamed for a number of years, but by 1913 it was called "Camp Roosevelt" — and a legend had taken firm root.[10]

Roosevelt Arch. The great basaltic arch dedicated by President Theodore Roosevelt on April 24, 1903, still welcomes visitors who enter Yellowstone by its north gate.

The legend of the "Roosevelt Tree" was based on the idea that "President Roosevelt slept here"; that he had camped under the shelter of the grizzled old Douglas fir tree, which still stands in front of the dining hall at Roosevelt Lodge, while vacationing in the Park in 1903. There is no doubt about where Roosevelt's campsites were in the Yellowstone Valley: the first was at the mouth of Blacktail Deer Creek, and the second was at the old Tower Fall Soldier Station, midway between Rainey Lake and the Overhanging Cliff.

Jack Haynes, the Park's photographic concessioner, opened a little store at Camp Roosevelt prior to World War I, and he said Howard Hays "brought in a lot of legend in connection with Camp Roosevelt and made quite a story out of it, and of course we believed all that stuff and made photographs of the Roosevelt Tree." That is, Jack believed it until one day when he was talking to Jim McBride, who had been with the Roosevelt party as an army scout. Jim said, "Jack, that Roosevelt Camp stuff up there is a lot of Hokum. I'll show you some day where the President camped."[11] That spoiled a good story, but the name has persisted in a slightly different form — as Roosevelt Lodge. After 1919, the various permanent camping concerns were reorganized into the Yellowstone Park Camps Company under the management of Howard Hays. The present lodge building was built the following year.

## HOWARD EATON

This early "dude-rancher," who conducted so many visitors through Yellowstone Park, is remembered in the name of the horse trail that circles the area.

Howard Eaton was not a born westerner; he came to the Little Missouri country of Dakota Territory from Pennsylvania in 1879 and established his Custer Trail Ranch five miles south of the town of Medora. For a time, Howard and his brothers prospered in the cattle business, but the terrible winter of 1886–1887 ruined that enterprise and caused the Eatons to convert their holdings to guest ranching, offering their many Eastern acquaintances a "'western experience' at a time when the true wildness of the frontier was rapidly disappearing."[12]

The ranch experience Howard Eaton was able to offer his well-to-do guests soon included more than the activities available on the home ranch (the ranch work of "real cowboys," horseback rides, and evening entertainments). Howard arranged trips to Standing Rock Indian Reservation, Custer Battlefield, and — particularly after the ranch headquarters were moved to Wolf Creek, west of Sheridan, Wyoming, in 1904 — short excursions in the Big Horn Mountains, longer ones in Jackson Hole, and extended excursions through Glacier and Yellowstone Parks. The magic of these outings can be sensed in the comment of guest Warren L. Hanna: "Howard Eaton always accompanied the party and personally took charge of its activities. He was more capable and energetic than any of his assistants, and his personality seemed to radiate through the entire caravan. With the pitching of camp, he became the life of the party, witty and refreshing, and with the coming of darkness and the gathering around the campfire, Howard always proved to be the most interesting raconteur of them all."[13]

Howard Eaton guided his caravans of horseback "dudes" on pack trips around a great loop of Yellowstone's trails in most of the seventeen seasons from 1905 until his death April 17, 1922. In so doing he established another form of visitor enjoyment of the Park's wilderness values, and dedication of the Howard Eaton Trail on July 19, 1923, was official recognition of his accomplishment.

## IMPERIAL

All the place names considered thus far have been personal, involving the names of people, but in this usage is a rarity — a name of visitor origin given to describe a feature.

It is probable that the powerful geyser now known as the Imperial began its activity in 1927, but there is no reliable record of eruptive behavior until July 14, 1928, when Ranger Roy G. Dale saw it in action. Dr. E. T. Allen from the Geophysical Laboratory of the Carnegie Institution visited the new geyser on July 27, and he was impressed as much by the volume of water discharged (60,000 gallons) as by the height (80–150 feet) and the duration (four to six hours).

Imperial Geyser. The dormant crater of the great geyser named by visitors in 1929 now appears as dead as the fire-blighted forest that surrounds it.

The word soon got around that the Park had another "big one" in the Lower Geyser Basin, immediately southeast of the Twin Buttes. An automobile road was opened to within a half mile of the new geyser so that the public could visit it; but what should the name be? By year's end there were a number of more or less appropriate suggestions: "Geyser of the Woods," "Great Geyser," "Twin Buttes Geyser," and "Valley Geyser."[14]

By the time the Park opened for the 1929 season, there was a new superintendent, and he perceived a publicity potential in the naming, duly reporting it thus: "The new geyser which broke out in the Lower Geyser Basin last year continued to play during the summer and was one of the main features of the park. This geyser was named by members of the National Editorial Association party while at Old Faithful on July 31. The name 'Imperial' was selected for the geyser."[15]

That was fine, except that the publicity was not warranted; the Imperial Geyser ceased performing two months later and was soon considered as defunct as the flashy Semi Centennial of 1922. However, it was not dead — only sleeping. After a dormancy of thirty-seven years, the Imperial returned to regal activity in August, 1966. The resurgence, a nearly constant jetting to heights of forty to seventy feet, did not last and Imperial is no longer "one of Yellowstone's major spouters."[16]

## SILVER TIP

While not exactly a Park place name, this usage is so inextricably tied into Yellowstone Park's history that it has to be considered here.

The original Silver Tip Ranch consisted of 157.2 acres of land adjacent to Yellowstone National Park in Slough Creek valley, where it hugged the north boundary line. G. Milton Ames took up a homestead there on June 16, 1913, and remained five years before selling out to two Yellowstone Park employees.[17] Ames had evidently added to his original entry, for the deed conveying his land to Burton C. "Bob" Lacombe (chief buffalo keeper) and Charles J. Smith (park ranger) on June 1, 1918, included two tracts totaling 296.93 acres.

The name by which the property was already known — Silver Tip Ranch — referred to the presence of grizzly bears in that area. It is said that Ames killed eight of those great bears (whose pelts are distinguished by white-tipped guard hairs) the first year of his homesteading, and that Joseph "Frenchy" Duret, the homesteader four miles up the valley, had taken 250 bears, grizzlies and blacks, during his residence on Slough Creek.[18]

By a strange coincidence, Superintendent Horace M. Albright and his assistant Joseph Joffe were escorting two partners of the House of Morgan, Thomas Cochran and William Hamilton, up Slough Creek to see the Lacombe-Smith property (which they were interested in buying) when Park Ranger Warren "Peck" Hutchings met them with word of the death of Frenchy Duret — killed June 12, 1922, by an enraged grizzly he had trapped. According to Joe, "We decided to go up to Frenchy's ranch. Mrs. Duret met us — cried on Mr. Albright's shoulder," and sold the ranch, lock, stock, and barrel, to the Morgan men, thus adding 416.08 acres to their purchase from Lacombe and Smith.[19]

Beginning in 1923, the Morgan partners turned the Silver Tip Ranch, which soon became the property of the Cornelius N. Bliss family, into a magnificent summer residence in a wilderness setting. The original ranch buildings were augmented through construction of a rustic lodge fronting on a polo field. This truly baronial hall, with its massive stone fireplace and great table split out of the trunk of a tree, was adorned with trophies of the chase. The furnishings — china, glassware, cutlery — were marked with the ranch insignia: a grizzly in a circle. As a remembrance of the bear-haunted early days, there was on the wall the No. 6 Newhouse bear trap that had failed to hold Frenchy's bear, and the Winchester Model 1895 rifle that had become only a broken club in the hands of the doomed trapper.

For many years the Silver Tip Ranch was both family retreat and guest ranch, dependent upon twenty miles of unimproved road for its access to the outside world; but with the abandonment of the road and establishment of a primitive area to the north, the ranch remains peculiarly isolated, its principal contact with the outside being the small aircraft that use the old polo field.

## DEMON/DEVIL

As previously noted, satanic nomenclature involving these words appeared in thirty-nine place names, with visitors apparently responsible for a little less than half. Of course most of their "devilish denomination" was wasted effort, for only one of their usages — and it is questionable — has survived the test of time.

An unidentified visitor writing under the pen name of "Emerald" in the *Omaha Republican* mentioned a twelve- by twenty-foot boiling sulphur spring at Sulphur Mountain, or Crater Hills, which he called the "Devil's Bath Tub."[20] This usage does not appear again, but the feature is probably the sulphur boiling spring visited by N. P. Langford and General Washburn on the morning of September 3, 1870 (it remains a sulphur-lined and very feisty hot spring).

Photographer Henry Bird Calfee named a feature in the geyser basins "Demon's Cave" in 1880, but exactly where it was located is not known. However, it *might* be the "Devil's Cave" noted in 1887 by the young lady known to us only as "Hattie" as being near Excelsior Geyser in the Midway Geyser Basin.

Judge H. N. Maguire used the name "Devil's Den" in referring to Dragons Mouth Spring in his lecture on "The New Eldorado" at Chicago's Farewell Hall on December 19, 1875. His presentation employed "photographs of nature, illuminated and magnified,"[21] which would indicate a very early slide lecture. This usage should not be confused with the Washburn party's "Devil's Den" on Cascade Creek, or the presently approved name on Tower Creek, above the fall.

"Devil's Half Acre" was the way a young lady with a fondness for devilish names described a thermal area she passed September 20, 1880.[22] It was the same place Superintendent Norris had named "Hell's Half Acre" (the present Midway Geyser Basin). Did Melvina perhaps have some Victorian scruple about the word "hell," so commonly associated with cursing? Her usage does seem to be a simple substitution.

An Iowa man by the name of Gardner S. Turrill made a wagon trip through the Park in the summer of 1898, and at the Mud Volcano he found that someone had written on the signboard "This is the Devil's Mush Pot." He noted, "I could scarcely repress a shudder at the fate of any person who might fall." There was no railing, yet some of his party ridiculed his estimate of the danger. Our cautious visitor — undoubtedly one of the many Christian Endeavor folk who passed through that summer — added: "As for me, I inwardly thanked God I was a Christian."[23] Turrill was not the only visitor to consign the Mud Volcano to the devil. W. S. Johnston called it the "Devil's Spring" (see the following entry).

Captain W. S. Johnston, who passed through the Park with a party in 1879, mentioned that the present Fountain Paint Pot was then called the "Devil's Paint Pot."[24] His usage was confirmed the next year by Melvina Lott.

Edward S. Parkinson published a book on his Yellowstone visit in 1892, and it provides a fine example of the ease with which popular place names get a bit mixed up. He mentioned two satanic usages on the upper terraces at Mammoth Hot Springs: The Devils Kitchen, one of the few such names to remain in use to the present (and the only one that *may* have been of visitor origin), and the "Devil's Parlor," of which nothing more is known than that it was nearby. His account is important because it contains an old story that is easily recognizable in a new locale:

> Back from the terraces, and farther up the mountains, are several openings or small caves. They have the euphonious names of 'The Devil's Parlor,' 'The Devil's Kitchen,' etc. They are extinct vents and are entered with safety by many of the tourists. The Devil's Kitchen is entered by means of a ladder about thirty feet in length. As you descend, the air gets warmer, and the odor of sulphur is easily distinguished. On one side is a natural cardrack, and many persons have left their names and addresses there. It is not many minutes before the perspiration starts and you have a desire to get out of the hole, and as there is nothing to prevent, you climb the ladder and soon breathe the cool air. On reaching the top the guide informed us that some years ago a hunter, whose cabin still stands in the gulch near the hotel, while walking over these terraces saw a pair of antlers protruding from the snow, and upon examining them found that an elk, in roaming the mountain, had broken through the crust of snow that had formed over the entrance to the kitchen and had hung there until dead.[25]

Of course, the foregoing tale is easily recognized as quite similar to the one by which the McCartney Cave, on the "Hotel Terrace," got its name (cf., Chapter 7).

A reminiscence of an 1883 visit to Yellowstone Park hints that the Devil's Kitchen may have had its name that early. Mrs. Hamilton of Bozeman, Montana, still had "a very keen memory of the Devil's Kitchen. We went down ladders into this deep cavern. I don't remember that we saw much by lantern light, but I remember the noise and the trembling — and the noise! and the smell! and the hissing, and the trembling, and the noise. We were just glad to get out from there to sunshine and firm ground."[26]

Among the many stories associated with the Devil's Kitchen, this one needs to be told again:

It is a deep black hole. A ladder is put down and all about the sides of the rock are the visiting cards of adventurous ladies. One of them had the delightful experience of meeting a big brown bear down there. She came up in a hurry, and when questioned as to the cause of her haste, she naively remarked, '*He* was at home,' referring to the accredited proprietor of the kitchen."[27]

In the heyday of this feature — the 1920s — there was even a refreshment stand nearby, appropriately styled "The Devil's Kitchenette"; but visitors are no longer allowed to go down into the solution passages underlying the hot spring terraces because of the danger of suffocation in carbonic acid gas. So, there is now no ladder into the depths of the Devils Kitchen, and the opening is securely fenced, but the name remains.

While speaking of devilish place names in this vicinity, I should mention that a wagon party from Butte tried to turn Old Bath Lake into devilish tableware in August 1885. They called that pleasant bathing place "The Devil's Soup Bowl," but the name was never important outside the fun-loving group that bestowed it.[28]

A visitor of 1905 who claims to have been with the first party to enter the Park from Cody, Wyoming, that year found that gangs of laborers still were shoveling snow out of the road at Sylvan Pass on July 5, below a "precipitous incline called

Devils Kitchen. The once popular labyrinth under the upper terrace at Mammoth Hot Springs is now closed in order to protect visitors from dangerous gases found there.

the 'Devil's Slide.'"[29] While Mr. Heath certainly was not the originator of this usage, he did lodge it firmly in the Park's literature. What he was referring to was the great slide of scree rock descending from Hoyt Peak, north of Sylvan Pass, and not the Devils Slide on Cinnabar Mountain below Gardiner, Montana.

The very hot thermal feature on Gibbon River now known as Beryl Spring was the "Devil's Washbowl" to the young lady we know only as "Hattie." She wrote to her aunt in Iowa: "One large pool called 'The Devil's Washbowl' is situated at the base of a mountain, the water jumps and boils, great volumes of steam arising that nearly hides the water. I think if he can bathe in that pool, he is quite a salamander."[30]

There is another of these devilish place names that requires an explanation. It is on the old road from the Lower Geyser Basin to Hayden Valley, where the track climbed up from Nez Perce Creek to Mary Lake — a wicked stretch known as the "Devil's Staircase" or "Devil's Stairway." A visitor of 1889 wrote that

> the road ascends Mary's Mountain, where the wagons must, for two miles, go up a stairway whose steps are from six inches to a foot and a half high and of solid rock. It is hard to imagine a road more difficult than this one for a team to draw a loaded wagon over. It would seem that Jordan is not the only road that is hard to travel. This hill-side is called '*The Devil's Stairway.*' Tradition fails to specify why this particular stairway became the property of his Satanic majesty. Just here it occurs that the Devil has altogether too large a share of the property in the Yellowstone Park to be to the credit of Uncle Sam.[31]

Mattoon did not know the origin of that name, nor do I; but the appropriateness is a matter of record. The stagecoaches bound for Lake and Canyon hotels went that way before the opening of the new route by way of Old Faithful and West Thumb (1892), and passengers had to get out and walk up that terrible grade. In late July 1890, Guy R. Pelton, a 64-year-old former New York legislator, descended from the coach with other passengers and began the hot, dusty tramp up the Devil's Stairway behind the slow-moving vehicle. But Mr. Pelton had a weak heart and the climb was too much for him; he collapsed and died by the side of the road.

Pelton's death came to the attention of the newspapers of our national capital, where the Park transportation business was portrayed as rapacious, with travelers "treated more like cattle than civilized people." An investigation was sought, with the result that politics took over. The transportation franchise was taken away from the Yellowstone Park Association (where it was in the hands of a Montana Democrat) and was given to the Yellowstone Park Transportation Company, organized by a Montana Republican. This arrangement did not improve travel conditions in the Park — it took better roads to accomplish that — but it did create a financial problem that plagued the concession operation for years. The hotel business was always a money loser, while the transportation did very well; operated under one management, the Park business balanced out, but dismembered it was never satisfactory. Blame it all on that "devilish" stairway.

It is necessary to add some names of this genre that are less certainly associated with the Park's visitation in order to be thorough. Mrs. Henely, an employee at Camp Roosevelt in 1913, had an opportunity to tour the Park that fall and she noted both a "Devil's Bump" and a "Devil's Ear" in the Upper Geyser Basin, which caused her to comment: "His Satanic Majesty seems to be much in evidence in this Park."[32]

Dr. Albert C. Peale of the 1871 Hayden Survey named a feature at the Washburn Springs the "Devil's Cauldron," but it is no longer identifiable. There was a "Devil's Gorge," which may have been a Norris name, along the Firehole River below Kepler Cascades, and the "Devil's Grotto"(1880) on the upper terraces at Mammoth Hot Springs certainly was one of his. The "Devil's Ink Pot" at the Washburn Springs probably received its name from Gooch & Whitfield in 1888 — from the color of its water. If Allen and Day are correct in identifying it with the largest of those nasty hot springs, then it is the present Inkpot Spring.[33] Historian H. M. Chittenden, in the 1903 edition of *The Yellowstone National Park*, pp. 216 & 322, places the "Devil's Inkstand" there also; therefore, it is not clear if these are different features or two names for the same feature. The Hoodoos below Terrace Mountain, sometimes called the "Limestone Hoodoos" to differentiate them from those at the head of Lamar River, were labeled the "Devil's Playground" on an undated lantern slide produced by photographer T. H. McAllister, but time has proven that people prefer to think the glaciers tumbled those great blocks down the mountain, not the devil. The "Devil's Punchbowl," a name that was never an approved usage, has now disappeared in favor of Punch Bowl Spring — which is what the Hayden Survey called it in 1878. As Chittenden once said, "It is some satisfaction to know that this rude and fiery nomenclature is gradually falling into disuse."

## MISCELLANY

William C. Gregg, the visitor who named the Ferris and Phillips forks of Bechler River, gave the Park a number of other place names that have stood the test of time. In his semiofficial investigation of the Cascade Corner (his name for the drainages of the Bechler and Fall Rivers), he enriched the Park map. Those names of Indian origin have been mentioned in Chapter 1 — (see Miscellany). The others are these:

- *Batchelder Column* (1921). A prominent rock on the east wall of Bechler Canyon, named for Amos G. Batchelder, executive chairman of the American Automobile Association, killed in an airplane accident near Washington, D.C., in 1920. He had been active in protecting the Cascade Corner.
- *Quiver Cascade* (1921). The only cascade on Phillips Fork of Bechler River.
- *Ranger Lake* (1921). A small lake on the westernmost edge of the Pitchstone Plateau named in honor of the Park's rangers. Mr. Gregg also listed a "Ranger Fall," but the name does not appear on the map or on approved lists, so what he had in mind is not clear.
- *Silver Scarf Falls* (1921). A cascade near Dunanda Falls but on a different branch of Boundary Creek. Of this name, which is descriptive, Gregg says: "Dr. Henry Van Dyke gave me the name of 'Silver Scarf Cascade' as his way of adopting one of our nameless beauties," and he adds, "I don't want to suggest a naming contest. I would take to my bed if an avalanche of 'Rainbow Falls' and 'Bridal Falls' came in."[34]
- *Sluiceway Falls* (1921). Another descriptive place name given a waterfall on the Ferris Fork of Bechler River.
- *Three River Junction* (1922). The place where the Phillips, Gregg, and Ferris forks unite to form Bechler River.
- *Treasure Island* (1921). An island in Bechler River near the lower end of Bechler Canyon, named in allusion to the locale of Robert Louis Stevenson's literary classic.

One of Gregg's place names, "Cavern Cascade," was renamed by him the following year (1922) as Tempe Cascade, presumably from an Indian word that could mean "cavern" or "pretty"; both have been offered as interpretations. In the magazine article cited in Note 1, Gregg seemed to regret the fact that the rules governing geographic nomenclature prevented him from naming something for the "present Secretary of the Interior, Albert B. Fall, who has entered the sanctuary of his park service with a truly National purpose and vision." He would have been appalled had he really understood the man and his motives, for Secretary Fall soon became the central figure in the Teapot Dome scandal, *and* the most determined enemy Yellowstone Park ever had.

Another visitor with semiofficial status was Dr. Herman C. Bumpus (1862–1943), who was chairman of the Educational Advisory Board of the National Park Service from 1931 to 1940. He proposed names for three Park features: Nymph Lake, a roadside feature north of Norris Junction (1931); the Ferric Caldron — for its iron-oxide coloring — in Norris Geyser Basin (1933); and Tuff Cliff on the right wall of Gibbon Canyon (1939). For his service to the Park in guiding the development of its system of trailside museums, Dr. Bumpus was honored in the naming of a prominent outcrop of volcanic rock in the vicinity of Tower Fall: Bumpus Butte.

A feature that probably received its name through common usage (which means we really don't know the origin) was given its earliest mention in the illustrated lecture of Mrs. O. W. Dean (1904). This little pool in Black Sand Basin, an area to the west of Old Faithful Geyser, displayed a pulsating action that would suck a handkerchief into the vent and then spew it out to be retrieved from the hot water with a stick. That became one of the things tourists were encouraged to do, particularly by the drivers of stagecoaches and buses. But the pool finally became sated with handkerchiefs and other things, and quit performing. The monthly report of Ranger Phillip Martindale for September 1927 contains this note: "Handkerchief pool functions very irregularly and takes many down that are not returned except in threads, one particle at a time." In 1950, enough trash was removed from the vent to restore some activity, but not for long. There now remains no trace of the once-popular little hot spring.

Handkerchief Pool was the feature Owen Wister was referring to when he remarked: "I always said, 'Let me assure you that I make it a rule to believe everything I hear.' But when they told me of a hole into which you could toss your soiled handkerchief and have it disappear and in a minute be thrown out washed, ironed, folded, and with a laundry mark, I drew the line."[35]

## NOTES

1. William C. Gregg, "The Cascade Corner of Yellowstone Park," *The Outlook*, November 23, 1921, p. 475.
2. The Ferris account was contained in an article that appeared anonymously in the Buffalo, NY, *Western Literary Messenger* July 13, 1842, under the title "Rocky Mountain Geysers." It was reprinted in that form by the Mormon newspaper, *The*

*Wasp*, Nauvoo, Ill., August 13, 1842, and, much later, the reprint was copied by the Helena, MT, *Weekly Independent* May 1, 1874, with a new title, "Visit to the Yellowstone Geysers 41 Years Ago." In 1900 the journal that was the basis of the foregoing accounts came to light. It was "Life in the Rocky Mountains," by Warren Angus Ferris. Dr. Paul C. Phillips put the journal into book form as *Life in the Rocky Mountains, 1830–1835* (Denver: Old West Pub. Co., 1940). The excerpt is from pp. 257–260.

3. Edwin J. Stanley, *Rambles in Wonderland, or, Up the Yellowstone and Among the Geysers and Curiosities of the National Park* (New York: D. Appleton and Co., 1878), p. 91.
4. George L. Henderson, *Yellowstone Park Manual* (1888), Mammoth Hot Springs, Yellowstone National Park, p. 2.
5. Hiram M. Chittenden, *The Yellowstone National Park*, ed. by Richard A. Bartlett (Norman: University of Oklahoma Press, 1964), p. 180.
6. S. P. Panton, "Early Days in Yellowstone Recalled by N. P. Surveyor," *Wolf Point News* (Mont.), November 23, 1939. The account of Lord Dunraven — Windham Thomas Wyndham-Quin, 4th Earl of Dunraven — was published as *The Great Divide: Travels in the Upper Yellowstone in the Summer of 1874* (London: Chatto and Windus, 1876). See pp. 180–346 for the visit to the Park.
7. "A Great Figure of the Yellowstone — General Hiram M. Chittenden," *Livingston* (MT) *Enterprise*, June 30, 1923.
8. "The Corner Stone Was Laid," *Gardiner* (Mont.) *Wonderland*, April 30, 1903.
9. John Pitcher, *Report of the Acting Superintendent of the Yellowstone National Park . . . 1905* (Washington: Government Printing Office, 1905), p. 3.
10. Louise M. Henely, *Letters From the Yellowstone* (Grinell, Iowa: Ray & Frisbie, Printers, 1914), pp. 12–13.
11. Anecdotes of Jack E. Haynes, recorded at Bozeman, MT, November 10, 1961. YNP audio tape No. 61–5.
12. Lee Silliman, "As Kind and Generous a Host as Ever Lived," *American West*, July/August 1979, p. 21.
13. Ibid., p. 22.
14. Horace M. Albright, *Annual Report for Yellowstone National Park, 1928.* Mimeographed, p. 17.
15. Roger W. Toll, [*Annual Report for Yellowstone National Park, 1929*]. Typed, p. 3.
16. T. Scott Bryan, *The Geysers of Yellowstone* (Boulder: Colorado Associated University Press, 1979), pp. 124–125.
17. A. A. Tony Bliss, *Silver Tip Ranch, 1922–1947* (privately printed, 1947), p. 1.
18. Doris Whithorn, "A Grizzly Got Frenchy," typed manuscript, 1963, p. 9.
19. Jack E. Haynes, Note recorded during an interview with Joseph Joffe, May 8, 1956. See Haynes file, Montana State University Library, Bozeman.
20. Reprinted as "Wonderland," Bozeman, MT, *Avant Courier*, January 1, 1881, p. 7.
21. H. N. Maguire, "The New Eldorado," Bozeman, MT, *Avant Courier*, Jan. 7, 1876, p. 1.
22. Melvina J. Lott, *Diary of a Yellowstone Park Trip, 1880,* typed MS in the collection of Montana State University, Bozeman, p. 4.
23. Gardner S. Turrill, *A Tale of the Yellowstone; or, In a Wagon Through Wyoming and Wonderland* (Jefferson, IA: G. S. Turrill Pub. Co., 1901), pp. 65–66.

24. "Tour Through Montana," Bozeman, MT, *Avant Courier*, October 23, 1879 (reprinted from the Poughkeepsie, NY, *Eagle*).
25. Edward S. Parkinson, *Wonderland; or, Twelve Weeks in and out of the United States* (Trenton, NJ: MacCrellish & Quigley, Printers, 1894), pp. 249–251.
26. Mrs. James M. Hamilton, "Through Yellowstone Park in 1883 with Mrs. James M. Hamilton," typed from an interview at the meeting of the Montana State University History Group, January 1967, p. 5.
27. Margaret Manton, *The Chicago Times*, December 6, 1891.
28. W. H. Dudley, *The National Park From the Hurricane Deck of a Cayuse, or, The Liederkranze Expedition to Geyserland* (Butte City, MT: Frederick Loeber, 1886), p. 39.
29. Charles A. Heath, *A Trial of a Trip; Notes From the Journal of Charles A. Heath* (Chicago: The Franklin Press, 1905), p. 42.
30. "Hattie," Letter written from Helena, MT, September 4, 1887, to Mrs. Mary Shulber, Dexter, Iowa. Copy in the Yellowstone Park Reference Library.
31. A. M. Mattoon, "The Yellowstone National Park, Summer of 1889," handwritten MS, revised and rewritten, 1907–1917, pp. 71–72. Yellowstone Park Reference Library.
32. Henely, *Letters* (1914), pp. 26 & 29; see Note 10 for citation.
33. Eugene T. Allen and Arthur L. Day, *Hot Springs of the Yellowstone National Park*, Carnegie Institution Pub. No. 466 (Washington: Government Printing Office, 1935), p. 412.
34. William C. Gregg, "The Cascade Corner of Yellowstone Park," *The Outlook*, November 23, 1921, p. 476.
35. Owen Wister, "Old Yellowstone Days," *Harper's Monthly*, 172 (March 1936), p. 475.

# 9

# THE ALLEY CATS AMONG THEM



Languages are, in one sense, living, growing things; adapting to changing conditions and needs by developing new words, idioms and meanings. English is particularly versatile in that regard because of its polyglot nature.

Out of that adaptability has come a type of place name which is colorful, expressive and generally transient, since it depends upon a shared remembrance. These informal names are seldom traceable to a particular origin, but seem to appear by spontaneous generation, with their persistence — and some are very long-lived — depending on an acceptance through common usage. Very probably, many more such names are given than survive, for they must be apt and useful to be accepted; if not, they are used a few times and forgotten, like a bad joke.

Yellowstone Park has its share of these informal place names — the "alley cats" of our nomenclature — several of which seem hardy enough to last forever, while others tremble on the verge of oblivion. Whichever, they have one thing in common; there is a story behind every one.

## SOAP HILL

Below Mammoth Hot Springs the road to the town of Gardiner, at the Park's north entrance, pitches steeply down a quarter-mile hill which is a terror for those who must go up to headquarters when driving conditions are bad; a place where the employee trying to get to work on a snowy morning (many live in Gardiner), or the resident trying to get home on an icy road after a shopping trip to town, was likely to fish-tail or spin-out — and end up stalled or in the ditch.

This devilish section of road will probably continue to be known locally as "Soap Hill," with its connotation of a very slippery place, forever, or until the grade is improved, whichever comes first!

Surprisingly, this informal place name antedates the automobile. Hiram M. Chittenden, the engineer officer in charge of roads and bridges in the Park, mentioned it in his annual report for 1901: "A very steep hill, called 'Soap Hill,' just below Fort Yellowstone has been cut out, replacing a 15 per cent grade with one of 8 per cent."[1]

Evidence of the earliest "Soap Hill" can still be seen as a raw scar on the old hot spring formation, climbing steeply from the present grade, on the right hand going up. That pitch must have been cruelly hard on horseflesh in any season!

Years ago, when Deate White was road foreman, he used to watch, in winter, for the arrival of a particular employee who lived at Gardiner but was a notoriously poor driver. If that man made it to work, there was no need to send out the sand truck. The road foreman used his rule-of-thumb very successfully for years, and I'm sure the patsy never knew what an important part his frequent struggles with "Soap Hill" played in the road maintenance program.

## BUFFALO FORD

Please understand that this crossing is not the one a short distance above the mouth of Lamar River, which got its approved name from the fact that the buffalo "show herd" was driven across the Lamar River there when moved from the Buffalo Ranch to the show corral at Mammoth Hot Springs, and vice versa, in the old days.

The informal place name we are interested in here is one that supplanted — hopefully, only temporarily — the much more appropriate name of Nez Perce Ford which was given in 1880 to the place in Hayden Valley where those embattled Indians crossed Yellowstone River while retreating through the Park in 1877. The substitution occurred in 1946, as the result of an accident which occurred there that winter.

It appears that a herd of buffalo tried to cross the river on ice which was not strong enough to support their weight. A ranger skiing down the river February 14, 1946, found the bodies of nine animals frozen in the ice near the ford, with other carcasses scattered along three miles of the river below that point. In all thirty-eight buffalo perished. When conditions permitted, a crew using a rowboat and a winch truck dragged the carcasses onto the shore where they were buried in a large pit excavated by a bulldozer. The burial site was a noisome place for some time as coyotes and grizzly bears persisted in digging up the rotting flesh.[2]

That tragedy gave rise to a tendency to refer to the place as "buffalo ford" rather than by its older, and more appropriate, name of Nez Perce Ford. The informal usage was reinforced when a picnic area was established on the river bank at the ford, with a National Park Service sign identifying it as "Buffalo Ford Picnic Area." A recent decision of the United States Board on Geographic Names has restricted the use of the Buffalo Ford name to its correct site on Lamar River, and confirmed the use of Nez Perce Ford for the Yellowstone River crossing in Hayden Valley.

## HAY'S/AFRICA/BEATTY'S LAKE

The road from Mammoth Hot Springs to the geyser basins climbs rapidly toward Golden Gate, traversing the rim of a brushy cirque part of the way. Directly below, and at least 250 feet down, is a small lake which has gone by several names — none of them official. The earliest name was "Hay's Lake," but why is uncertain (though it seems to involve concessioner Howard Hays). "Africa Lake" was used during the

Beatty's Lake. An unofficial memorial to truck driver Roy Beatty, who rode a runaway vehicle into the water June 16, 1931. Other names for this body of water are Hay's Lake and Africa Lake.

1920s and 1930s, particularly by bus drivers, because its shape, as seen from the road, does resemble the outline of that continent. More is known about the third name, "Beatty's Lake."

Roy Beatty was a local man who worked in the Park as a truck driver. He came from a family which had homesteaded close to the north boundary in the early days (Superintendent Norris had allowed his grandfather, James, to run cattle in Lamar Valley, where quite a few were eaten by the Nez Perce Indians as they fled through the Park). Roy was a good-natured giant of a man, and proud of his strength which he liked to demonstrate by picking up the shop anvil with one finger in the swage hole.

On June 16, 1931, Roy was coming down grade in his truck when he lost control of it above the lake, and was soon off the road. He could have bailed-out at that point, but he didn't; he rode it down through the brush and into the lake. The truck was in bad shape but Roy was all right, and from then on, as long as there was anybody around the Park who remembered, the lake was his.

## LAKE LACOMBE

Burton C. "Bob" Lacombe was one of the men who made up that original ranger force which took over protection of Yellowstone National Park from the Army on October 1, 1916. He is best described as a hard, practical man "whose toughness had considerable depth." He became Chief Buffalo Keeper at the ranch in Lamar Valley in the 1920s, where he and another old timer, Joe Douglas, kept bachelor hall. Both men liked to "celebrate" as well as any, and, when Ben Arnold invited them to Tower Ranger Station for a bit of Christmas cheer, they went, though the weather was bitterly cold.

Sometime in the early hours of the morning Bob decided he ought to go back to Lamar, so, he went out, cranked up his old Buick and started for the ranch alone. In that day the road went about as now until it past Junction Butte, then it swung more to the north and came down to a small lake where it turned to the right and passed along the shore. Drifted snow had obscured the road near the lake and Bob missed the turn; he kept straight on, about fifty feet on the ice of the lake, until it broke under the weight of his vehicle.

The water was not very deep, but Bob got thoroughly wet reaching shore through the shattered ice, and, as he was "well-oiled," his situation was more perilous than it might otherwise have been. There was nothing he could do but stagger back to Tower, where he arrived nearer to death than a man has a right to be and live (in the language of his peers, "they gave him a couple more and he survived.")

Now, about the "Lake Lacombe" name, Wayne Replogle has this to say:

> In 1932, I rode the motorcycle patrol, such as it was, out of Tower. Evil [Eval] Knievel wouldn't have had a chance with what I had [to ride] in those days. Anyway, I noticed a rude sign out in the middle of the little lake which had on it 'Lake Lacombe.' It seems that the boys had thought enough of the incident to give it the name, since their illustrious pal had such a struggle there. The sign remained another year or two.[3]

## CHINAMAN'S GARDEN/GULCH

The strip of meadow crowded between Gardner River and the hill on the west, just below the upper crossing on the North Entrance road, doesn't look like much of a garden, but before the river carted off so much of the good ground, it was a fine one. Superintendent Norris, who put in a half-acre of potatoes, turnips and other hardy vegetables there in 1879, said of his gardening that it was "far more successful in production than in protection from vandalism at a place so remote and hidden from our block-house."[4]

As was the custom at frontier Army posts, the soldiers stationed at Mammoth Hot Springs (where the post was first called Camp Sheridan, and later Fort Yellowstone) raised a "post garden" to supplement their very plain rations. A map of the Park road system prepared by the Corps of Engineers in 1892 shows a "Mil. Post Garden" where Norris had planted, and a wagon visitor of 1898 has left an interesting note on how at least some of the produce was used:

> Just before making the halt, we came to a government truck patch guarded by a soldier. I asked if the garden 'sass' was for sale. After looking both ways to see that no one was watching, the faithful guard agreed to let us have what we wanted. Accordingly we purchased a sack full of cabbages and turnips, and the soldier was fifty cents better off financially. Our part in this transaction was scarcely honest.[5]

It appears that the Army gave up the post garden some time soon after the turn of the century, and the site was taken over by the

Chinaman's Garden. An unofficial place name that remembers Sam Wo, the last gardener to work the old truck patch beside the North Entrance road. Earlier it was the Fort Yellowstone post garden.

Yellowstone Park Association as a good place to raise fresh produce for its hotel dining rooms. A sketch drawn in 1909 by architect Robert C. Reamer, for a "Chinaman's Garden cottage,"[6] indicates that Sam Wo had been there long enough to be identified with the garden, where he probably lived in the old guard shack. A long-time employee of the hotel company, "Pete" Hallin, remembered helping to build Sam Wo's house in 1917 in accordance with the Reamer design. The result, as the Haynes photograph made in October of that year shows, was a delightful little building with a "Hansel and Gretel" touch to it.

Sam Wo was well liked, with a local reputation as a cook and healer, as well as gardener. It was considered a real treat when he would cook for the employees mess at Mammoth, and he knew how to cure stiff necks and headaches. Mr. Ambrose, who drove the express between Gardiner and the hotel at Mammoth, used to stop to pick up vegetables for the kitchen, and, if he had some ice cream in his load, he would open one of the ice-packs and let Sam take a bowl of it; which was reciprocated with a round of Chinese wine.

In 1922 Sam Wo gave up gardening in the Park and went to Spokane, Washington, but his fifteen years beside the North Entrance road are remembered in a surprisingly tenacious place name — Chinaman's Garden.

Another Chinaman, Sam Toi, ran a hand-laundry in the old McCartney cabin in the mouth of Clematis Gulch (the building which had been the Park's first hotel); that is, he did until the building was destroyed by fire on December 4, 1912. For some years thereafter Clematis Gulch was known locally as "Chinaman's Gulch."

It is no coincidence that both these Chinamen, Wo and Toi, had "Sam" for a first name, or at least were so-known. The American frontiersman preferred a "handle" to a proper name any time, hence the "Yankee Jim," "Little Gus" and "Society Red" approach to personal identification; it was the sure mark of acceptance. For a Chinaman there was only one handle — it was Sam, which cut neatly across all the linguistic difficulty of a foreign speech.

## MCGUIRK'S

Matthew McGuirk was put out of the Park too soon to acquire status as a hotel man; he remains just a squatter — an interesting, "lucky Pierre" sort of man.

Early in August, 1871, McGuirk began to develop a site at Hot River, or "Boiling River," where the subterranean runoff from Mammoth Hot Springs emptied into Gardner River. A Col. J. D. Chestnut, of Bozeman, had camped there with a party of invalids the previous month, creating a short-lived tent camp called "Chestnutville." Encouraged by their claims of nearly miraculous cures, McGuirk erected a cabin, built bathing pools and put in a garden on the little flat close above.

In the course of that work he brought in the first wheeled vehicle ever to enter what is now Yellowstone National Park. His ox cart had to be taken apart at the "Second Canyon," now Yankee Jim Canyon, so that it and the freight could be packed over a very dangerous trail. The trail was not the only danger at that time, for Sioux war parties made frequent forays against the Crow Indians, whose reservation was then in the Yellowstone Valley between the "first" and "second" canyons, and they were just as willing to attack small parties of whites.

On March 9, 1872, McGuirk claimed "Not to exceed 160 acres" on the banks of the Gardner River, but he was nine days too late; the land had been withdrawn from the Public Domain through the establishment of Yellowstone National Park on March 1. McGuirk petitioned Secretary of the Interior Columbus Delano for a lease on what he called "McGuirk's Medicinal Springs," but it was not granted and Superintendent Langford had him put out of the Park in 1874.

The buildings were used as quarters for government employees until 1889, when the acting superintendent, Captain Moses Harris, had them torn down. McGuirk's name never found lodgement on the Park map and will soon be as thoroughly forgotten as his paltry improvements.

## SPLIT ROCK

The rock is there yet, a mighty block of sandstone tumbled from the bluff above Eagle Nest Rock into the Gardner River where the frosts of many years have cleft it neatly through. It is in plain view of the present road, yet few who now pass by know its name, and fewer still have heard of the bizarre event which took place there in the park's early days.

The first road through the Gardner Canyon was on the west side of the river at that point, passing close against the "Split Rock," which was ideally situated for what two shifty characters from the town of Gardiner had in mind for the evening of July 4, 1887.

The "brain" (what a loose usage!) behind this caper was an unemployed teamster, William James, who was unemployed because a detail of soldiers had caught him poaching beaver in Gibbon River the previous winter. He had been put out of the Park, and on the blacklist as far as further employment there was concerned, and was hanging around the town of Gardiner in the hope of getting back the equipment confiscated at the time of his expulsion. The other man was a local nobody by the name of Charley Higgenbotham, and what they had in mind was nothing less than robbery of the Army paymaster.

The paymaster was to arrive on the afternoon train to Gardiner, on his monthly round to pay the soldiers stationed in the Park (always in cash in those days). Of course the two highwaymen expected the paymaster would take the evening stage to Mammoth Hot Springs, so they planned to halt the coach at the Split Rock and get the paymaster's black satchel.

Accordingly, the robbers hid themselves behind the Split Rock, and while they were waiting for the stage, a buggy passed by with two men in it. The driver was Ed Lamartine, a road foreman employed by the Army engineers, and he was taking the paymaster to the post (Camp Sheridan); but the robbers didn't know that.

In due time the stagecoach *Bighorn* came up and driver Thea Hamm was halted by two masked men brandishing pistols. It was a very amateurish holdup which netted the robbers only sixteen dollars

Split Rock. The unofficial name of a massive boulder across Gardner River from the Eagle Nest Rock. Standing on the edge of the old road, it hid the men who bungled the Park's first stagecoach robbery on July 4, 1887.

and two souvenir pocket pieces taken from a passenger, Judge John F. Lacey of Oskaloosa, Iowa. It was a crime which might have gone unpunished, for the robbers were not identified at the time, except that James took such a fancy to those foreign coins — a French "Napoleon" of 1811, and an Italian Lira with the head of Pius IX on it — that he couldn't resist bragging to friends in the Gardiner saloons. He and Higgenbotham were soon betrayed by those they consorted with and were gathered in by the soldiers.

Both men got prison terms, so that James not only had the honor of being the first man expelled from the Park by the Army for a misdemeanor (poaching), but also the first sent to prison by the Army for a felony (robbery). But there was some ultimate good out of all this.

The passenger whose souvenir coins led to apprehension of the robbers later went to Congress as a representative from Iowa, and it is poetic justice that, when the way was at last clear for the passage of legislation which would give Yellowstone National Park effective legal machinery, it was Representative Lacey, a man who had been robbed in the Park, who introduced H.R. 6442 — "an act to protect the birds and animals in Yellowstone National Park, and to punish crimes in said park, and for other purposes."[7] We know it as the Lacey Act.

## BUFFALO RANCH

The place shown on current maps of the Park as Lamar Ranger Station used to be the Buffalo Ranch, a name which was an approved usage until the government decided to quit handling its tame herd as if they were domestic cattle. But, "Buffalo Ranch" lingers on as an informal designation for the collection of log buildings and massive corrals which are the remaining evidences of a failed enterprise based on a noble idea.

For the genesis of the idea, it is necessary to understand what was happening in and around Yellowstone Park when Philetus W. Norris assumed the superintendency (1877). The great herds of wild buffalo which once numbered in the millions on the high plains of the West had been reduced to a few scattered bands and there was no real hope for their survival. Within the Park the remaining 300 to 400 "mountain bison," a somewhat smaller, darker sub-species adapted to the northern Rocky Mountains, were being decimated by hunters whose activity was not even forbidden by Park regulations until 1883. Under those conditions, Norris could see only extermination for the American bison — unless they could be saved by domestication.

The Lamar Valley with its broad meadows, protected by lofty peaks and ridges, and its remoteness from the principal routes of travel, seemed ideally suited to animal husbandry, hence his recommendation that "There two or three spirited, intelligent herdsmen might (in addition to profitably rearing domestic animals) also thoroughly protect and, by capture of the young, gradually domesticate any desired number of them [buffalo]."[8]

Norris was motivated by an apocalyptic view of a day when the buffalo would be "elsewhere unknown, save in the natural histories of extinct species," and it was his vision that "they might here prove a perpetual attraction to the eye, under proper regulations."[9] Possibly as a beginning toward his idea of domestication of buffalo calves, Norris induced a stockman, James C. Beatty, to put some cattle in Lamar Valley (there has to be a source of milk to raise buffalo calves); but that approach failed when hostile Nez Perce Indians, retreating through the Park in 1877, slaughtered the cattle for travelling rations.

In his report for 1879, Norris argued the "necessity for systematic and permanent protection of all, and domestication of some of the most rare and valuable of animals in the eastern portion of the park by the assistance of lessees of hotel and ranch sites."[10] Fortunately, nothing came of that idea beyond authorization to employ an energetic frontiersman, "Rocky Mountain" Harry Yount, as resident gamekeeper, but Harry, after a winter spent in a lonely cabin near the mouth of Soda Butte Creek, came to the conclusion that the job was more than one man could handle and resigned — with the recommendation that what was really needed was a "reliable police force." The gamekeeper position was used for a few more years without accomplishing a thing.

Shortly after the turn of the century, there was a renewed interest in the old idea of ranching buffalo in Lamar Valley. This was the result of alarm over the decline of the Park's buffalo, which were viewed by fledgling conservation interests as the last wild herd in the United States. This effort to "save" the buffalo resulted in establishment of a tame herd based on a ranch operation at Rose Creek in Lamar Valley. It was a success as a means of insuring an adequate number of buffalo, but also a biological disaster in that the inevitable cross-breeding between imported plains buffalo and the local mountain buffalo resulted in hybrid animals not really typical of either strain.

The ranching operation was terminated following World War II as not consistent with National Park Service policy, but, at its peak, it was large. The work on the ranch was supervised by a chief buffalo keeper who had an assistant, several irrigators and other help as the

Buffalo Ranch. The former name of the present Lamar Ranger Station. Before World War II, it was dedicated to "saving" the Park buffalo herd.

season required. The irrigated meadow lands along Lamar River produced 800 to 900 tons of hay which was fed to the buffalo during the bad weather, and there were facilities, such as corrals, chutes and sheds, necessary to the inoculating, castrating and slaughtering which was part of the maintenance of a healthy herd of about 800 animals.

What remains at Rose Creek is only part of the old "Buffalo Ranch," but it is a poignant reminder of an idea which seemed good — it just didn't turn out right!

## SEPULCHER/KITE HILL

The high, barren hill immediately north of Mammoth Hot Springs was first known as "Sepulcher Hill" from the early civilian cemetery established on its summit; but that informal usage was never popular and has given way to "Kite Hill," which is descriptive of its present usefulness.

The cemetery was established June 10, 1883, with the burial of Mary J. Foster, who remains an unknown beyond the fact that she was 33 years old and was born in Madison Co., NC. An item in a local newspaper a year and a half later states: "Out of seven occupants of Sepulcher Hill at the Springs three deaths are fairly due to the baneful influence of intoxicating drink."[11]

There is little doubt the old cemetery was essentially a "boot hill" for the Park and the gateway community at Gardiner. Of the seven burials for which cause of death is known, there was one murder, two suicides, two drownings and a winter death in a snowslide. There are evidences of thirteen burials, the last in 1897, though the cemetery was little-used after the Army took over protection of the Park in 1886 (the Army established its own burial ground south of Capitol Hill, and the Gardiner townspeople opened their "Tinker's Cemetery" about the same time).

Perhaps the local disinterest in the Sepulcher Hill name is due as much to the fact that there is a prominent feature nearby which is officially known as Sepulcher Mountain (so-named by the Barlow Party of explorers in 1871 because the south end of its ridge had the appearance of a tomb, as seen from Mammoth Hot Springs), and two similar names in the same vicinity was *too* confusing. Anyhow, "Kite Hill" sounds better as a place for the kids to be on a windy day!

Sepulcher Hill. An unofficial name for the small hill immediately north of Mammoth Hot Springs, deriving from the early graveyard on its summit.

## MISCELLANY

Perhaps, in gathering up these odds-and-ends of informal nomenclature, it is as well to begin with the "Tinker's Cemetery." This burial ground a mile west of Gardiner, Montana, on the old county road, was outside the Park at the time of its establishment, but is nominally within the boundary as revised by the act of May 26, 1926.[12] According to local tradition, the cemetery got its name from the fact that the first burial there was of an itinerant mender of pots and pans, crushed under his wagon while greasing a wheel. The cemetery holds the graves of some well-known Park characters: "Uncle" John Yancey, James McBride, "Geyser Bob" Edgar, "Little Gus" Smitzer, and the sand painter, Andrew Wald.

The Gardiner addition to the Park provided another informal place name — "Game Ranch." This usage refers specifically to the vicinity of the old ranch house which now serves as quarters for Park employees. It seems to have developed out of an idea that the land added to the Park to provide winter range for the antelope was a "ranch" similar to the Buffalo Ranch, but it was never operated in that manner. Though not appropriate, the name has stuck.

A triangle of land lying between the Gardner and Yellowstone rivers to the foot of Mount Everts is known as "the rifle range" from its former use in the training of soldiers stationed at Fort Yellowstone. Evidences of that use remain in the target butts built in front of a small hill. The area also was used as a golf course — by government employees — during the 1920s and 1930s.

Half way to Mammoth Hot Springs from Gardiner, the road makes a tight turn around a swampy draw with a small thicket in it. That is the site of "Whiskey Spring," a place where the soldier trudging back to Fort Yellowstone after a spree in the Gardiner saloons could get a much-needed drink of water. Incidentally, the five miles back to the post was a death march for two men: Private Andren Preiber's death was noted as from "exposure near Gardiner," while Private Frank Monaghan "drowned in Gardner River about one mile from Gardiner, while intoxicated."[13]

Over the years, a greater number of informal place names have attached to features at Mammoth Hot Springs than anywhere else in the Park because of the year-around occupancy of a headquarters settlement there. Many belong entirely to the limbo of the past, as "McCartney Creek" (now Clematis Creek), the "Pioneer Hotel" or "Job's Hotel" (James C. McCartney's cabin), "Chinaman's Gulch" (the draw from which Clematis Creek flows; so-called in the days when Sam Toy ran a Chinese laundry there); the "Beehive" (one of the old Camp Sheridan buildings notorious for crowded tenancy), the "Bucket-of-Blood" (a stagecoach driver's dormitory where a particularly sanguinary encounter took place), and the "Milk Ranch" on the hill above the hotel (a dairy; necessary in those days before adequate refrigeration). Then there were concession enclaves as "Henderson Town," "Clark Town," and "Haynes Town," and the government's own "Norris Blockhouse"; gone and quite thoroughly forgotten.

A few of the localisms have persisted. "Lost Lake," still identifies a small pond below the old power house; a scummy body of water in the residence area which should not be confused with the real Lost

Lake near Tower Junction (the one at Mammoth suffers from a lack of fresh water since the power house was shut down). Likewise, the double street with the landscaped center remains "The Esplanade," but note, this is not the feature on the Mammoth terraces officially known by that name. The little log building south of The Esplanade, at its lower end, was built in 1897 by A. T. French, mail contractor on the Cooke City route, for his quarters, but Victor Hjelm, a painter for the Yellowstone Park Company, occupied it for so many years that the name "Hjelm House" became firmly attached. The Fort Yellowstone building which houses the community library, gym and meeting rooms is still "The Canteen," a hold-over from Army days when it was the post exchange. There is one more building with a distinctive appellation; the square structure of gray stone, with the green tile roof — the old Park headquarters, at the upper end of The Esplanade — known to several generations of rangers as "the temple of truth," which certainly fits its faintly Oriental appearance.

Continuing on from Mammoth Hot Springs, along the road toward Norris Junction, a pond below the old reservoir — barely visible from the road — was known for many years as "Joffe Lake" because it was the favorite fishing place of Joseph "Joe" Joffe, who was assistant to the superintendent from 1922 to 1960.

Above the "Seven-mile Bridge" about three-quarters of a mile, on the bank of Gardner River, was the "Meat Ranch." It was another of those facilities made necessary by the primitive refrigeration of an earlier day (beef cattle, from a herd grazed in Gardners Hole, were slaughtered there for the hotels, all of which were within a night's run by the meat wagon).

A half-mile north of Clearwater Springs was "Brickyard Hill," a name still familiar to a few old timers. The old road, which went over a low ridge instead of around it as now, took its unofficial name from the fact that the Army used clay obtained there to make bricks for the chimneys of the Camp Sheridan buildings (the antiquity of this usage is attested by records of the United States Engineer Department).[14]

About a mile from the old Norris Soldier Station, the road used in Model-T days kept to the east of the present alignment, passing along the edge of a swampy area which no way resembles a proper lake. However, it was known to road crew men as "Colitus Lake" from some obscure association with Park Engineer Lord.

On the Norris Cutoff, that road which is the center section of the figure-eight system, there are two formerly important features which have unofficial names. A short distance below the Virginia Cascade was "Devil's Elbow," a place where the old road in the canyon bottom made a hairpin turn around a prominent boss of rock. This feature can be seen by looking directly down from the present scenic road at a point about 1000 feet west of the cascade; the name probably is more than a hundred years old, as it was well enough established to appear in the 1890 edition of the *Haynes Guide*. The other name, "Blanding Hill," is at least equally old. It was named for James Blanding, one of Lt. Dan Kingman's crew foremen on the initial construction work in 1886. The grade by which Blanding carried the road up from Virginia Meadows onto Solfatara Plateau remained in use until the cutoff was realigned as a Mission–66 project in 1957. It is remembered by old Park hands as the first place to become impassible on the cutoff as winter snows closed in.

An unofficial usage which appeared on the West Entrance road about the turn of the century was "Christmas Tree Park," which aptly described the new forest coming in on the old burn along the first seven miles within the Park boundary. Time changed that young forest into a stand of mature lodgepole pines so that the name gradually lost its significance (that is what happened to an earlier use of the name by the Geological Survey in its attempt to describe a similar forest at the head of Winter Creek in the Gallatin Range).[15]

Strangely, the southern reaches of the Park have not been burdened with as many unofficial place names as its northern verge, which could be a result of largely transient use by persons with few ties to the past; also, the features in the geyser basins were already so well supplied with approved place names that there was no real need for others.

An exception to the foregoing statement is found in the two sites of former hotels in the Lower Geyser Basin: the "Marshall House" at the mouth of Nez Perce Creek, and "Fountain Hotel" near the Fountain Geyser complex. Early in 1884 George W. Marshall was given a lease allowing him to build a frame hotel in the angle between Nez Perce Creek and Firehole River (replacing his earlier stage station on the other side of the river), and that hotel — the Marshall House — became the first hostel in the geyser basins, with a considerable settlement around it. The construction in 1890 of Fountain Hotel two miles to the south and adjacent to important thermal features, led to abandonment of the earlier site (a fate which also came to Fountain Hotel in 1917). The visible evidences of both these places caused them to retain landmark value long after their usefulness as hotel sites had ended.

Of course, the Old Faithful area cannot be dismissed without noting the "Fossil Deer Tracks," a place in a bend of Firehole River, just above the upper highway bridge, which was important mainly as a place for employee wiener roasts, and the old "Lunch Counter for Bears." That early bear show at the garbage dump on the edge of a rudimentary automobile campground flourished during the 1920s, with a ranger on horseback lecturing each evening to the visitors seated on logs in front of the platform where the bears fed on hotel slops; that is, they did until a big black bear chased a little black bear though the assembled crowd, thus putting an end to the Old Faithful bear show.

The road from Old Faithful to West Thumb Junction has two unofficial place names related to stagecoach robberies in the Park. One, the "Turtle Rock," is known only from Jack Haynes' story of the 1908 holdup; except for that, the usage would have vanished even more completely than the abandoned road it once was a landmark on.

This forgotten place alongside the crooked old road up Spring Creek, east of the Upper Geyser Basin, was the locale of what has been called the largest stagecoach robbery of the twentieth century (debatable on some grounds), but, if the number of vehicles involved is a fair measure, it certainly was. From that viewpoint it may even have been the largest holdup in the whole brief era of staging in the West, though the loot would hardly have interested a "Black Bart."

The armed robbery on August 24, 1908, of 149 tourists being transported in seventeen vehicles out of a convoy protected by an armed cavalryman, was an audacious undertaking — and also a very successful one! The then Park Superintendent, General S.B.M. Young, had suffered through a previous stagecoach holdup, and, though he eventually solved that one, the recriminations had steeled his determination never to let *that* happen again. His answer to the ever-present threat of robbery by highwaymen able to use any convenient site along several hundred miles of wilderness road, was to require all stagecoaches of the operating companies to move in convoys protected by soldiers.

Thus, the 25 vehicles which moved out from Old Faithful Inn that morning were preceded by the usual armed escort. But a holdup man had stationed himself in Spring Creek Canyon, about four and one-half miles from the Upper Geyser Basin, and he wisely hid behind that rock Jack Haynes calls the Turtle Rock[16] until the soldier and the first eight vehicles had passed. Since they were travelling a hundred yards apart because of the dust, he had put a half mile between himself and the protector of the slow-moving column before stepping into the road with his rifle.

Taking a young man out of the first coach he stopped (to keep that driver from moving on to spread the alarm), he approached each vehicle as it came to a halt and commanded the occupants to throw their money and other valuables into the sack his hostage held out. The "take" was not great; a reported $1,320.95 in cash, and jewelry estimated at $760.75, but that does not include non-negotiable drafts and tickets totaling $10,030.00. A careful examination of the "List of Victims and Their Losses" published immediately after the robbery shows only seventy-one of the 149 suffered a loss; evidently the others refused to pitch anything into the bag, and the lone bandit was not able to enforce his demand when they held out on him — one of the problems of attempting such a big job single-handed.

Who was this bold highwayman? The best guess is that he was William Binkley, a fugitive who had escaped from the guardhouse at Fort Yellowstone where he was serving a sentence for poaching in the Park. The description fit him and he had been heard to remark "that it would be an easy matter to hold up the coaches and that he would 'show them' when he got out."[17] Anyhow, if it hadn't been for this robbery, the Turtle Rock would have slipped quietly into oblivion without leaving a literary ripple.

The other holdup took place at "Shoshone Point," a place with a view of part of Shoshone Lake. This robbery of July 29, 1914, was not as big an affair as the one at Turtle Rock; a mere 15 coaches and 164 passengers. It was also less successful, since the robber — Ed Trafton — was soon caught and sent to prison. The Shoshone Point name has remained in use from at least as early as 1896 down to the present.

Just before West Thumb junction is reached, a curve in the road exposes a magnificent view eastward across Yellowstone Lake. This viewpoint received its unofficial name of "Lake View" from photographer F. J. Haynes in 1896. It is a place where one can share the appreciation of the explorers of 1869 for the serene vista which they described in these words:

> As we were about departing on our homeward trip we ascended the summit of a neighboring hill and took a final look at Yellowstone Lake. Nestled among the forest-crowned hills which bounded our vision lay this inland sea, its crystal waves dancing and sparkling in the sunlight as if laughing with joy for their

> wild freedom. It is a scene of transcendent beauty which has been viewed by but few white men, and we felt glad to have looked upon it before its primeval solitude should be broken by the crowds of pleasure seekers which at no distant day will throng its shores.[18]

Before automobiles were admitted to Yellowstone Park there was a short cut on the road from West Thumb to Lake Hotel, by means of which traffic wishing to avoid some of the heavy pulling along the shore of Yellowstone Lake might pass over the ridge from Arnica Creek to Bridge Bay. That road went through an area of forest variously known as the "Knotted Pines" or "Knotted Forest." The road is no longer open to public travel, but the forest is still there — and still of some interest because the striking deformities used as architectural trim in the construction of Old Faithful Inn came from that locality.

The East Entrance road has two features with unofficial names which require mention. One is "Teton Point," a place 1.5 miles west of Cub Creek from which the peaks of the Teton Mountains can be seen 60 miles to the south (this name was carried in every edition of *Haynes Guide* from 1930 on). The other usage — really two names — described an unusual road structure on the east side of Sylvan Pass. The pitch of the ravine up which the road was constructed became so steep near the top that it was necessary to build a spiral bridge in order to gain the required elevation and yet keep the grade within reasonable limits. The original (1903) structure was a wooden trestle known as the "Spiral Bridge," or "Corkscrew Bridge"; later rebuilt in concrete, the abandoned overpass can still be seen by anyone looking down from the shoulder of the present road. There used to be a "Corkscrew Hill" on the old road between Craig Pass and DeLacy Creek, but it didn't deserve the name as it was only a series of steep S-curves.

Continuing on the Grand Loop from Lake Junction, there is a curious meander of Trout Creek, immediately above the road, known by several names over the years. It came to public attention indirectly, as "Trade Mark Creek,"[19] and was called "Trout Cr. Trademark" on a Haynes postcard which appeared in 1916. The 1930 and subsequent editions of *Haynes Guide* referred to it as "The Monad" in the index, but in the text only mentioned that Trout Creek's meander had the shape of a Chinese Monad. These usages are related in that the logo of the Northern Pacific Railroad — the linked red and black commas — was taken from the design which formed the centerpiece of the old

Korean flag. It was a symbolic rendering of the Chinese idea of "yin and yang"; a dualistic philosophy which sees everything in terms of opposites which are yet complementary, as, male and female, good and evil, light and dark.

There are two unofficial place names near the road as it approaches the Grand Canyon area. "Spurgeon's Beaver Slide" is a place on the mountain side, about 1.3 miles south of Chittenden Bridge, where Captain W. F. Spurgeon lowered Army supply wagons down a steep slope with ropes during General Howard's pursuit of the hostile Nez Perce Indians in 1877. For many years, a sign at a roadside pullout kept the name current, but removal of that marker after World War II in order to reduce traffic congestion had the effect of consigning the usage to limbo.

The other name, also nearly forgotten, was applied to the place where the "bear show" was reestablished after it became a too dangerous exhibition in its original setting adjacent to the Old Faithful automobile campground. The new location was on Otter Creek, at the end of a half-mile spur road, where a "Bear Feeding Ground," consisting of rustic seats behind chain-link fencing overlooking a concrete platform, was completed in 1930.[20] By 1936, this had become "The Grizzly Bear Show," and the site was well established as "The Grizzly Bear Grounds" when the facilities were demolished in 1947 (in accordance with changes in Park policy with regard to animal "shows").

Skipping across the Washburn Range to the Lamar Valley, a few unofficial place names in that area are worth mentioning. At the mouth of Soda Butte Creek, Lamar River flows against the toe of a ridge where rock has been blasted away to make room for the road. In earlier times — before the turn of the century — the miner's road to Cooke City passed over that ridge by a sidehill cut known as "Jackson's Grade," for Sam Jackson, an early squatter at Rose Creek where the "Buffalo Ranch" was later established. Jackson also built a bridge over Lamar River near the present picnic area as part of a toll road scheme, but his eviction by Superintendent Robert E. Carpenter, along with other squatters, in 1884 put an end to his plans. A faint scar still visible above the present road marks the line of Jackson's Grade.

Just over a mile west of the bridge over Lamar River, there is a small knob close to the road, on the north side. That open, windy place was the site of a Civilian Conservation Corps camp in the 1930s. It was such an inhospitable place, early and late season, that the CCC boys called it "Little America," from a feeling of kinship with Admiral

Peary's party in Antarctica. The Park road crew later used the site as a mix pit and traces of that operation remain.

On the Yellowstone River, at the mouth of Yancey Creek, is a famous fishing place known locally as "Yancey's Hole." It was there that Uncle John Yancey sent the guests who stopped over at his Pleasant Valley Hotel for some fishing fun, and it was there that he caught the fine river trout which were regular fare at his table. There is a tendency, of late, to confuse the Yancey's Hole site on the river with Yancey's old hotel and ranch site in Pleasant Valley, but they are not the same.

The last of these unofficial place names is the "Sand Spring" on the road to Tower Junction (about five miles from Mammoth Hot Springs and close below the road). This fine little spring boils out of a sandy bottom, and several generations of residents at Park headquarters were so firmly convinced its water was better for them than the domestic supply piped down from Gardner River that they made regular trips to the spring for bottles of that good water. It might have been better for them in former years as the potable water tested a bit high in fluoride, but the present danger of contamination of such roadside springs by giardia, certainly outweighs any other advantages they might have now. Incidentally, this spring should not be confused with the Sand Spring on Witch Creek in Heart Lake Geyser Basin, a feature named by the Jones Party in 1873.

## NOTES

1. H. M. Chittenden, "Annual Reports Upon the Construction, Repair, and Maintenance of Roads and Bridges in the Yellowstone National Park. . .," *Annual Report of the Chief of Engineers for 1901* (Washington: Government Printing Office, 1901), Appendix FFF, p. 3784.
2. Mildred Ericson, "Animal Accidents in Geyserland," *Nature Magazine*, 41/6 (Jun.–Jul. 1948), pp. 317–318.
3. Wayne Replogle, "Rip' Rap," *Yellowstone News*, August 7, 1975.
4. Philetus W. Norris, *Report Upon the Yellowstone National Park to the Secretary of the Interior. . .1879* (Washington: Government Printing Office, 1880), p. 4.
5. Gardner S. Stilson, *A Tale of the Yellowstone; or, In a Wagon Through Western Wyoming and Wonderland* (Jefferson, IA: G. S. Turrill Pub. Co., 1901), p. 79.
6. Former Yellowstone Park Company plans deposited with the Montana Historical Society, Helena; see hanger No. 3, plan No. 7.
7. Aubrey L. Haines, *The Yellowstone Story* (Boulder, Colorado: Associated University Press, 1977), II. See index for more information on this event, and for the text of the Lacey Act.
8. Philetus W. Norris, *Report Upon the Yellowstone National Park. . . 1877* (Washington: Government Printing Office, 1877), p. 843.

9. Loc. cit.
10. Norris, *Report . . . 1879* (1880), p. 21. See Note 4 for citation.
11. "Park Notes" Livingston, MT, *Enterprise*, Nov. 15, 1884, p.3.
12. Haines, *The Yellowstone Story*, II, pp. 331–332; see Note 7 for citation.
13. "Record Book of Interments, Post Cemetery, Fort Yellowstone, Wyo.," Yellowstone Archives, Doc. 155, pp. 2 & 5.
14. Lieut. Wm. E. Craighill, "Report of Lieutenant William E. Craighill, Corps of Engineers," *Annual Report of the Chief of Engineers for 1890* (Washington: Government Printing Office, 1890), Appendix EEE, p. 3596. The location of the "brick yard" is shown on a map of the United States Engineer Department, File 334 [Route maps of Yellowstone Park roads, 1891], 26 blueprints dated July 9, 1892. See sheet No. 9.
15. United States Geological Survey, *Geological Atlas of the United States — Yellowstone National Park Folio, Wyoming* (Washington, D.C., 1896), Gallatin Quadrangle sheet.
16. *Yellowstone Stage Holdups* (Bozeman, Montana: Haynes Studios, 1959), p. 15. This material also appeared as "Yellowstone's Stage Holdups," in the 1952 *Brand Book* of the Denver, Colorado, Westerners.
17. Haines, *The Yellowstone Story*, II, p. 153; see Note 7 for citation.
18. Charles W. Cook, David E. Folsom and William Peterson, *The Valley of the Upper Yellowstone*, ed. by Aubrey L. Haines (Norman: University of Oklahoma Press, 1965), p. 40. The entry is for September 29.
19. Reau Campbell, *Campbell's New Revised Guide and Descriptive Book of the Yellowstone Park* (Chicago: Rogers & Smith Co., 1914), p. 118.
20. Roger W. Toll, "Annual Report for Yellowstone National Park, 1931" (mimeographed), p. 13.

# 10

# THOSE THAT DIDN'T MAKE IT

*Mount Dawes & Mount Pomeroy*
*Arthur's Seat*
*Wedded Trees*
*KKK Tree*
*Anthony Falls*
*Chestnutville*
*Norris Sinks*
*Bob Ingersoll Geyser/Spring/Terrace*
*Lost Lake*
*Miscellany*

Throughout the history of Yellowstone National Park there have been place names used that no longer can be related to a definite place. These are names that have lost currency, and in falling into disuse have become so dim in our collective memory as to remain mere ghosts of usage past, relegated to a limbo where they are neither fully known nor quite forgotten. Most often this happens because of the disappearance of the people who knew the where and why. However, the process of forgetfulness is greatly aided by natural changes such as those resulting from fire, flood, landslide, and vegetative growth; and by the activities of man in constructing, abandoning, or even obliterating cultural works. Such place names remain literary enigmas, known mainly from vague references in older records.

George L. Henderson, early assistant superintendent turned hotel proprietor, appears to have been responsible for more of the "lost" place names than any other person. This was a result of his enthusiastic but not always appropriate naming and renaming of features that were mentioned in his published guide without enough information to pin them down. But Henderson was not alone in that; other diarists and writers of the pioneer period used place names carelessly.

Some of those lost place names, faintly remembered but no longer identifiable as to place, are the subject of this chapter.

## MOUNT DAWES/MOUNT POMEROY

The Scottish poet Robert Burns had something to say about "the best-laid schemes o' mice an' men" that often go astray, and that certainly held for the Hayden Survey's plan to name mountains after the men who had been particularly helpful to the survey and to the new national park.

The field parties returned to the Park in 1872 with a list of such names to be bestowed upon unnamed peaks. An entry in the diary of William Blackmore, an English anthropologist along as a guest, noted these preferences: Clagett, for William H. Clagett, the delegate to Congress from Montana Territory; Pomeroy, for Samuel C. Pomeroy, United States senator from Kansas; Anthony, for Henry B. Anthony, United States senator from Rhode Island; Edmunds, for George F. Edmunds, United States senator from Vermont; Dawes, for Henry L. Dawes, United States senator from Massachusetts; Trumbull, for Lyman Trumbull, United States senator from Illinois; Delano, for Columbus Delano, secretary of the interior under President Grant; Grant, for President Ulysses S. Grant; Sheridan, for Lieutenant General Philip H. Sheridan, and Sherman, for William T. Sherman, then General of the Army.[1]

General Sheridan had already been honored by the attachment of his name to a summit in the Park — Mount Sheridan, which received its name from the Barlow party in 1871 — and only four of the others were ever so remembered (all much later).[2] On the basis of map information, it would seem that Blackmore's list was no more than idle speculation; but this is not so.

A member of the Hayden Survey's first or northern division, Joseph Savage, makes this statement in a letter sent to his hometown newspaper August 13, 1872. After mentioning that he had just visited "Mt. Stevenson and Langford on the south and Mt. Pomeroy and Dawes (just named) on the southeast," he added, "The remaining peaks as yet not christened."[3] The failure to enter this information on the map of that summer's work probably was a result of the death of Chief Topographer Adolf Burck before the field notes could be plotted. That work had to be done in Washington, D.C., by personnel unfamiliar with the season's labors, and it is altogether likely that some of Burck's data were lost or overlooked.

The men who would have been honored had the Mount Dawes and Mount Pomeroy names survived were eminently deserving of that recognition. Henry L. Dawes was an outstanding veteran of the U.S. Congress where, beginning in 1861, he served eight consecutive terms in the House, followed by three in the Senate. It was said of him that there was very little lawmaking from 1861 to 1893 in which he was not consulted. His influence became dominant in appropriation matters, and he was responsible for the establishment of the U.S. Fish Commission, our national weather service, and Indian education. To that should also be added that he fathered our system of national parks in the sense that his Yosemite Grant legislation (1864) became the model for Yellowstone's organic act in 1872. Concerning his influence on the latter, Cramton has this to say: "While his hand may not have written down the words, he undoubtedly shaped the lines of the draft and was the power behind the scenes that made things move."[4]

Samuel C. Pomeroy did not have the influence of Dawes in the Senate, but as chairman of the Committee on Public Lands, he had the honor of introducing S. 392 and shepherding it through the legislative toils. It was his measure, and not Delegate Clagett's companion bill H. R. 764, that was signed into law by President Ulysses S. Grant on March 1, 1872, accomplishing our first reservation of wild lands "for the benefit and enjoyment of the people" *under Federal management.*[5]

The attempt in the summer of 1872 to honor Dawes and Pomeroy for their very great accomplishment came to nought, and it is a peculiar irony that we do not even know which among the many great peaks southeast of Yellowstone Lake were selected to be their memorials.

## CHESTNUTVILLE

The location of this earliest settlement in the vicinity of Mammoth Hot Springs is lost in the sense that the site is no longer recognizable because of the blurring effect of subsequent use of the area around Hot River.

The dramatic rescue of Truman C. Everts, the unfortunate member of the Washburn party of Yellowstone explorers lost for thirty-seven days in the fall of 1870, introduced several Bozeman and Fort Ellis men to the hot springs on Gardner River. What they saw was the possibility of developing spas based on the presumed curative value of the hot waters, and that idea led to the establishment of three settlements in 1871.

The earliest of these, and the least permanent, was a rude gathering somewhere in the vicinity of Hot River. It came into existence prior to June 22, at which time it was visited and described by a correspondent of a Helena newspaper, who wrote:

> About 3 o'clock we hove in sight of the camp on Warm Spring Creek [the present Gardner River], and met with a hearty reception from the denizens of that impromptu village called

> Chestnutville. Colonel Chestnut,[6] the founder of this embryo city, "called us," We "saw him," went in and partook of a lucious feast of any amount of elk, trout, calf, side dishes. . . . At dusk we gathered around a huge camp fire, fill our pipes, and stretching ourselves on the ground listen to the wonderful description of the country beyond, of curiosities and sights, many of which the Washburn party are ignorant of, but all these did not astonish us as much as some sights now before us. To explain, one of the party, Mr. Ben Green, of Bozeman, who has been an invalid for years, was in the circle, talking and laughing with great gusto, apparently in the full enjoyment of health. In answer to a query of ours he said, "Why man the springs did it. There is nothing like them. I am going to winter here."
>
> Mr. John A. Lechener, of Helena, who came to the springs by the wise advice of Dr. Glick, is gaining his former good health rapidly, and feels confident that he will be himself again in a few months. Col. Chestnut is loud in the praise of the waters of the springs, while Dan Houstin and others all invalids, will not budge an inch from the declaration that they are the best springs in America for helping the poor in health.[7]

The correspondent also noted that Harry Horr and James C. McCartney were already at work erecting their house in the mouth of Clematis Gulch, though they did not formally claim the site until July 5, 1871. Matthew McGuirk, the third settler, began his improvements along Gardner River above the outlet of Hot River in November of the same year, by which time Col. Chesnut's establishment seems to have disappeared; perhaps they were the same, with only the management changing. But that is at best just a guess.

## ARTHUR'S SEAT

Chester A. Arthur, twenty-first president of the United States, was the first chief executive to visit Yellowstone National Park. We also know that he was not in the best of health that summer of 1883, so it is hardly surprising that he "sat down to rest after climbing the terraces," at a point that "commands a splendid view of the Mammoth Springs."[8] For a few years thereafter the soldier boys who guarded the fretted beauty of the terraces from souvenir hunters were able to show tourists the very place where the President had sat, but that "seat of Government" can no longer be located exactly. The best guess at this late date is that it was somewhere along the outer lip of The Esplanade, where the view drops precipitously to the "Hotel Terrace" far below.

## NORRIS SINKS

This is another of the Henderson place names that cannot now be related to a particular place, though the "unfathomable slime" he described seems to have been somewhere between the Monarch and Vixen Geysers. The following is his account of an experience there:

> Here the writer on one occasion was driving from the Monarch in a heavy shower, the whole basin being veiled in a dense fog. My companions were Dr. and Mrs. J. W. Joyce of Cincinnati and Miss Lambie of St. Paul. I took the wrong side of the pine ridge and to my consternation found myself entering the sinks. Old Dan, one of the best animals that ever wore a collar in human service, sank to his knees and, with a groan of despair, lay over on his right side. The other had as yet good footing, but was so terrified that his restless motion was gradually working the horrid mud underneath into soft mush. I sprang from the carriage and held the fallen horse firmly to prevent all ineffectual struggles. The two ladies were soon on safe ground among the trees and the doctor, with a coolness that did him credit, managed to loosen the tugs and clear the harness. The nigh horse was instantly out of danger. In the meantime I had sunk to the knees and knew that when the next thin crust gave way a bottomless abyss was all that I could expect. The doctor held the tongue aloft and my chief danger now was in being caught by Old Dan when he should make his leap to life. The dumb brute seemed to have reasoned it all out. With a cautious movement he raised his head and seeing where Dave stood in safety he wisely aimed for that locality, and in doing so resorted to the only possible way to save me, for had he made a single forward movement of two feet we would have inevitably disappeared and perished together. The crust that held me up had also sustained the fore part of his body and with a movement, impossible to describe, he crawled a few feet on his knees where finding the substance firm, he was able to drag his hind legs out of the slime, making a snoring sound as the ooze flowed into the space made by his body. There was joy indeed and a clapping of hands as the noble beast stood once more on solid footing. All this

> transpired so quickly that it was impossible to realize how little time it had taken. With a picket rope attached to the hind axle the carriage was extricated and we were soon standing beside the Vixen.[9]

## WEDDED TREES

There were two such landmarks along the Park's roads; one on the Norris Cutoff, between the five- and six-mile posts from the Canyon, and the other near Sylvan Lake on the East Entrance road.

The Wedded Trees on the Cutoff received the most notice — as well as the earliest. In his 1896 guide book (p. 29), F. J. Haynes noted the "Twin Trees" as standing on the south side of the road, 5.5 miles east of Norris Junction, or exactly halfway through the cutoff. A "sagebrusher" of 1898 — this refers to a visitor using his own or a hired rig, rather than a "tourist" traveling by public transportation — mentioned that they

> soon came to a lot of sulphur beds, emitting a cloud of steam. Near the road we observed a natural curiosity in the shape of two big pines, linked together by a limb, which was firmly grown to each tree. A sign-board informed us that these were the "Wedded Trees." There is no other freak of this description in the Park [not so, but he didn't know that]. Some wag had written beneath the name on the sign-board these words: "What God hath joined together, let no man put asunder."[10]

This feature was mentioned in guidebooks for nineteen years, but was dropped from notice after the 1915 edition, possibly because of a rerouting of the road to accommodate automobile use about that time.

The other Wedded Trees — near Sylvan Lake — were first noticed, at least as far as available records admit, by Alma White in September 1919. She did not comment other than that they were "at the left," or south of the road.[11] The 1921 edition of *Haynes Guide* (p. 30) picked up this feature, but with an equal disinterest — adding only that it was 8.1 miles from Lake Junction as the road went then. Editions after 1923 do not list these Wedded Trees.

## BOB INGERSOLL GEYSER/ TERRACE/SPRING

George L. Henderson's attachment of Ingersoll's name to three Park features has created a triple enigma, for not one is identifiable nowadays. But in the 1880s, these thermal features were described in the newspaper-like *Yellowstone Park Manual* published for the benefit of tourists.[12]

Robert Greene Ingersoll was an American political leader famous for both oratory *and* an agnostic viewpoint. He visited the Park in 1884 and was toured about by Henderson, who appears to have been so taken by his guest's "genial and kindly manner" that he honored him by bestowing his name on a terrace and spring — No. 52 in either edition of the *Guide* — at Mammoth (a feature described only as having "maintained its vigor and beauty while many of its neighbors have died out," which is not very specific), and a geyser on Geyser Hill in the Upper Geyser Basin. The Bob Ingersoll Geyser, "like its namesake, is a clear, sparkling fountain and, though its eruptions are irregular, they are steadily gaining in strength and volume" (p. 4, 1885 ed.). That is nice to know, but it doesn't locate anything, and it doesn't explain a strange statement made thirty-four years later.

In 1913 a young lady enroute to her place of employment in the Park stopped overnight at the Wylie Way camp on Gibbon River. Something she heard or saw at that camp caused her to remark later, "It is an oft repeated story that the night Bob Ingersoll died every geyser in the Park went into violent eruption."[13] He died in 1899 — fourteen years earlier — so there *must* be an interesting story behind that statement. Or is that, too, lost forever?

## THE KKK TREE

On the old Norris Cutoff road, about a mile east of the top of Blanding Hill, there used to be a grim reminder of one of Yellowstone's winter tragedies. Eight feet north of the roadway stood a pine tree that had, carved into its trunk, the letters "KKK," and below that a cross and three notches. It marked the place where three employees of the Yellowstone Park Hotel Company died the night of November 28, 1923.

A White stake-body truck had gone from Mammoth Hot Springs to Canyon Hotel with supplies for the winter-keeper, smashing its way through snowdrifts that had already closed the road to general traffic. On the return that night, the truck stalled in the snow and the four men decided to stay with it. They kept the engine running, took up floorboards to get heat from the exhaust pipe — the temperature was about twenty degrees below zero — and arranged themselves for a miserable night. Three men were in the cab and a fourth was in his bedroll in the back of the truck.

The men in the cab had covered their heads and the upper parts of their bodies with a blanket, and sometime in the night it caught fire from the hot exhaust pipe. That awakened Jim McDevitt, the man sleeping in the rear, and when he opened the cab doors, Ed Wright fell out from one side and Dale Soroka from the other. They were dead from exhaust gases, as was the third man in the middle of the seat.

A road improvement project in 1960 called for widening the roadway at that point. The KKK Tree was found to be in the way and the feasibility of removing it was taken up with Superintendent Lon Garrison. He decided the tree had to go and ordered that the inscribed section of trunk should be sawed out and saved. That grisly memento was lost somewhere, along with all knowledge of exactly where it once stood.[14]

## LOST LAKE

This usage is not related to the approved place name given to the small lake one-half mile above Roosevelt Lodge, nor to the unofficial name of the weedy pond that is the remaining vestige of a former fish hatchery near the old powerhouse at Mammoth Hot Springs. Instead, it refers to a feature shown on the topographical maps of the Geological Survey as an honest-to-goodness feature from 1885 until 1961.

As noted in Chapter 5 (Miscellany), this nonexistent feature resulted from persistent attempts to place a body of water, variously called "Two Water Lake" (Dunraven, 1874) and "Two Ocean Pond" (Norris, 1879) on or near the Continental Divide approximately three miles nearly west of the old West Thumb junction. The aerial survey upon which the 1961 edition of the Park-wide map was founded proved beyond a doubt that no such lake existed, and yet its existence was strongly supported at various times in the past.

Dr. Fedinand V. Hayden's statement that the lake he saw and on the shore of which he camped in 1871 was "about two miles long and one wide"[15] is no more believable than another statement of his to the effect that Bridger Lake simply didn't exist. I was involved in field surveys in that area for several seasons, in connection with the Grant Village development, and can testify that there are no lakes on the divide between Shoshone Lake and the thumb of Lake Yellowstone, nor are there any natural pockets in that terrain to basin a lake of such an extent.

In past years, much literature, which I prefer to class as legend if not outright hoax, has appeared concerning this "Lost Lake." The inquiries of fishermen baffled in their attempts to locate the lake led Ranger-Naturalist Trusten Peery to question old-timers as to why such a sizeable body of water could not be found. This is what he learned:

> Some light was thrown on the subject by Dick White, our general foreman, in a conversation with the author last year [1936]. Dick reports that a road foreman by the name of Billy Ferrell found the lake in the summer of 1915. Ferrell had lost some horses near Cold Springs, about seven miles from West Thumb on the way to Snake River. In his search for his horses, Ferrell traveled northwest and joined the Old Faithful–West Thumb road about a mile west of the crossing of the continental divide. He found the mythical lake about three miles from the latter road, and, having some fishing tackle with him, tried his luck with splendid success. A bit more information was obtained this summer from Jack Mears who worked in Yellowstone Park for many years as a packer. Jack says he often visited Lost Lake to fish, and that the lake is located only three miles southwest of West Thumb. His last visit to the lake was in 1921 and this is the last record, to my knowledge, of anyone having seen the lake.[16]

Lee Whittlesey adds further testimony that would indicate the existence of a lake, but he did not document the source of his information. He noted that "Ranger Frank R. Oberhansley set out to find the lake on August 8, 1938. He took four and one half hours to find it and reported that it drained into Shoshone Lake. Nonetheless, the lake does not show up in aerial photographs. It is possible that Lost Lake, like other glacial lakes, has filled up with vegetation."[17] I suggest the foregoing should be taken, if at all, *Cum grano salis!*

## ANTHONY FALLS

Though Hayden's Geological Survey of the Territories had planned to name something in 1872 — probably a mountain — for Henry B. Anthony, U.S. senator from Rhode Island, nothing came of that proposal.

Forty-nine years later, the name Anthony was again suggested as suitable for perpetuation on the map of Yellowstone National Park, but *not* in commemoration of the senator who, when the legislation to establish Yellowstone as a national park was under consideration, January 30, 1872, took a determined stand against using the English game preserve (really a hunting reserve idea) as the model for the new area. Instead, William C. Gregg, who was in touch with Mrs. Alice B. Winters of the General Federation of Women's Clubs, was inclined to use the name for one of the unnamed waterfalls of the "Cascade Corner" as a tribute to Susan B. Anthony (American feminist and leader of the women's suffrage movement). Though he was "sure we will have an 'Anthony Falls' that will be a splendid monument . . . to a grand woman,"[18] nothing more was heard of his proposal.

## MISCELLANY

It was mentioned earlier that more "lost" names can be traced to George L. Henderson than to any other person. That statement rests on the fact that his names were catalogued in the two newspaper-like guides he published for sale to tourists (see Note 12), and were thus perpetuated in the Park's literature; however, his descriptions were mostly so inadequate that many of the features cannot now be identified with certainty. Common usage might have "saved" more of his place names had they caught on with the visiting public, but they were generally faddish — sometimes drawn from the classical or romantic literature then in vogue, often quite trivial, and always exceedingly imaginative.

Of the ninety-two place names Henderson gave to features that are no longer identifiable, twenty-seven were at Mammoth Hot Springs or its immediate vicinity, and forty-four were in the Lower Geyser Basin, both areas where the Henderson family operated hotels. There were ten in the Norris Geyser Basin, nine in the Upper Geyser Basin, and one each in the Monument and Middle geyser basins.

Since there is very little to be said about place names that cannot be related to present Park features, the names are listed here under their general area, with only such comments as tend to preserve the intent and flavor of Henderson's nomenclature. Quotations used are his.

Mammoth Hot Springs and vicinity:

*Abysmal Lake.*

*Caldron Lake.*

*Castle Rock*, in Gardner Canyon.

*Cellingham Crag*, on the face of Terrace Mountain: "Forty or more people can stand on this crag, but as each one steps from the mainland to this island in mid air, you shiver in fear that the 'last feather' may send it away into the wilderness where a million other crags lie in broken fragments."

*Chamber of Snorers*. "It is amusing to hear the lady visitors tell which of these melodious tones resembles her husband's night song."

*Cold Lakes*, at the foot of *Jacob's Ladder Terrace*. Henderson considered this presently nonexistent water supply the reservoir for a "wonderful mechanical process" whereby water was heated and elevated to a *Pulsating Geyser* (also now missing) above the terrace.

*Crimson Lake*. This *might* be the present Painted Pool. It is possible that there were more pools (Henderson's "lakes") on the upper terraces before the Yellowstone Park Improvement Company cut the original forest between Pinyon Terrace and Snow Pass to produce the lumber with which the National Hotel was built.

*Dover Plateau*, likened to the "cliffs of Dover, hence its name"; but this is *not* Terrace Mountain, which comes nearer to such a comparison, because Henderson specifically noted that feature as *Limestone Mountain*.

*East Gardner Lake*. Then, as now, there were several ponds along the lower course of Glen Creek, making it impossible to decide which was meant.

*Giant's Stairway*. This feature and the nearby *Infant's Stairway* appear to have been on the old tree-covered slope (*Cedar Terraces*) below the Mammoth Hot Springs settlement proper.

*Hermit's Cave*, somewhere above the present Stygian Cave, was described as the "subject of one of Haynes's admirable stereos in which a hermit is seen sitting bareheaded with streams of hot water descending the almost perpendicular walls above his head."

*Jelly Cave*, noted as having a sufficiently direct connection to *Boiling River* (now Hot River) that "clouds of vapor" from that subterranean flood rose from it. This was noted as explorable for a long distance, "when the wind blows into the mouth of the cave." No such feature exists now, and there is no possibility that Henderson was speaking of the McCartney Cave, since he mentioned it separately.

*Lone Rock, The*, on the east side of the unidentifiable *Dover Plateau*.

*Mammoth Cave*, mentioned as opening below the present White Elephant Back Terrace, but the fate of this feature is also something of a mystery.

*McMinn's Spring*, which evidently broke out in 1884 somewhere above the White Elephant Back Terrace, was touted for its curative properties — specifically for curing the "chronic rheumatism" of Mrs. Silas McMinn, wife of a settler on Reese Creek north of the Park.

*Oriental Terrace.* Henderson's remark, "where most of the fine coating is now done," indicates one of the active, lower springs — Minerva or Cleopatra. Whichever it was, that is also the locality of *Abysmal Lake.*

*Sentinel Rock*, in Gardner Canyon, between Eagle Nest Rock and his *Castle Rock*, is described as a "human-like, upturned face gazing at the heavens with a fixed and stony stare"; it may have become a casualty of later road building in the canyon.

*Snow Cave*, south of *Ladies' Lake* and near *Crimson Cave,* was also presumed to expel poisonous gases.

*Spittoon, The.* This feature, "hidden in a small valley," was imaginatively coupled to another — *The Three Zebras.* One was a "Spitting zebra" that expelled drops of water into the spittoon.

*Sulphur Pits, The.* Pockets containing crystals of sulphur were found in the area from Orange Mound Spring to *Mammoth Cave*, and that condensate from escaping gases was considered evidence that "beds of sulphur underlie this whole valley" — hardly a supportable supposition!

*Teller Plateau*, named for Henry M. Teller, Secretary of Interior (1882), might be the present Prospect Terrace.

*Twin Cones* were noted east of Orange Mound Spring. "The jelly on the west cone palpitates periodically as if it were a sobbing child. The throat of the other looks like flesh and discharges tepid water in a most singular manner, as if it vomited." Enough?

*Two Crescents, The.* Something between Kingman's wooden trestle and Rustic Falls led Henderson to ruminate about "'sermons in stone and lectures in running brooks.' And the truth that man and man's work is nature's chief end and glory." Whatever glory this feature represented appears to have been largely destroyed by subsequent road building.

*Wizzard's Hat Terrace.* Cheops Mound is approximately right in basal dimension, but it does not support forest trees, as Henderson mentioned.

Lower Geyser Basin:

The "lost" names Henderson gave to features in the Lower Geyser Basin, where most thermal features already *had* well-established names, are summarized by groups in a pamphlet he published in 1891.[19] Beginning with what is now the Fountain Group, he states:

> The Yellowstone Park Association will have completed before the season of 1891 opens a splendid hotel near the Fountain Geyser. At this point may be seen the Parnassus Basin [from Greek mythology; a mountain sacred to Apollo and the Muses] with its Evangeline, Little Jewel, Raven, Fritchie's Well, Star Spangled Banner, the Clock [probably Clepsydra Geyser], Hiawatha, and Thanatopsis, also in close proximity . . . the beautifully terraced Museum, the Mammoth Paint-Pots [now Fountain Paint Pot], and the great Green Globe Lake, the only known boiling lake on the planet that contains uric acid.[20]

Henderson suggested that one thousand dollars would open a carriage road giving access to his *Ebony Basin,* the area around Hot Lake, a mile east of Fountain Hotel; and to his *Mammoth Geyser Basin*, surrounding the present Great Fountain Geyser. The advantages of the proposed road (which *was* subsequently built) were listed as:

> In one day the tourist may see Othello Fountain, Walpurgia Lake, Black Warrior Geyser, with its alabaster and ebony rings, discharging its boiling, never-ceasing flood into the steam-enveloped Walpurgia, day and night, summer and winter, without a moments suspension. Near Ebony Canyon there is the Champaign, Marguerite, Faust, and a host of other geysers, named and anonymous, also the Parlor Basin with its half dozen beautiful boiling fountains. The Venus and near her the Inominata, two subjects worthy of a day's study by painter or poet.
>
> The Mammoth Fountain [Great Fountain Geyser], whose terraces are the most perfect in their coral-like structures, and whose discharges constitute a liquid pyramid of over 300 feet from base to peak. Near this great geyser is the Seraph, her terraces and creek, the splendor of which surpasses the wildest dreams of the paradises of the poets; also the mysterious Blistered Terraces, terminating at the Sepulchre Geyser [earlier, Sepulchre Spring, and *possibly*, present Buffalo Pool], and the

> equally mysterious Trinity, known as the Fairy, Fiend, and Fury.
> The ancient Hecla is to be seen not far from the Mammoth Fountain, as it sends up vast volumes of steam with occasional jets of water. Visitors with time and taste might spend a week most profitably in the Ebony and Mammoth Basins.[21]

Another carriage road Henderson thought necessary to give hotel guests access to the west side of the Lower Geyser Basin would have crossed Fountain Flats to the old Freight Road (called *National Park Avenue* then), diverging from that route up Fairy Creek to the thermal area at the foot of Twin Buttes. Fortunately, the road was not built as proposed; but had it been, this would have been the result, as Henderson saw it:

> I shall name but a few of the wonders that will be met with on this tour of a day or a week of days, as inspiration might dictate. On either side of a road to the Brilliant may be seen Lindren's Garden in Chemical Valley, Microcosm Basin, The Nimbus, Alladin's Cave, Giant's Foot-prints, Maiden's Coffin, Caldrons, Minor and Major, Golgotha Geyser, Ragged Edge, Bleeding Heart, Loch Gelt, Loch Cloud, and Loch Fine, Sea Wave Geyser, and Specimen Terrace at the foot of Lone Butte, Queen's Laundry [now Queens Laundry], and near it Domes Angelo and Raphael, Hour Geyser, The Puffer, Golden Trap, Steam Roarer, and a host of other geysers, paint pots, steam vents, and gas-aqueous geysers.[22]

In the other geyser basins, where he had no hotel business, Henderson was less active in naming features. At Norris his failures were Crystallization Lake, Dancing Dunes, Fairy River, Leap Frog Lake, Millstone of God (earlier, just The Millstones), Sulphur Spray, Whirlpool, White & Black Recording Terrace, and White Paint Pots. In the Upper Geyser Basin his "lost" names, in addition to the Bob Ingersoll Geyser mentioned earlier, were Green Glacial, Subterranean Geyser, Maelstrom, Three Graces, Triplets, and Volcanic Moses — all in what is now inaccurately called Biscuit Basin (which Henderson seemed to think he discovered).

A name George L. Henderson gave a thermal feature in the Lower Geyser Basin in 1885 provides a fitting close for this chapter. He called it Inominata, probably from the Latin *inominatus*, or unlucky — which will certainly do splendidly for whatever it was!

## NOTES

1. William Blackmore, personal diary No. 7, 1872, p. 104 (photocopy in the Yellowstone Park Reference Library).
2. The Geological Survey used the name "Sherman Volcano" (1888) for the feature now referred to as the Mount Washburn Caldera. Clagett Butte (1926) was named by Park Superintendent Albright; Grant Peak (1931) was named by Cadastral Engineer Bandy, and Mount Cowan, well outside the Park, was named by the Geological Survey.
3. Joseph Savage, "The Wonders of the Yellowstone," Lawrence, Kansas, *Western Home Journal*, September 19, 1872.
4. Louis C. Cramton, *Early History of Yellowstone National Park and its Relation to National Park Policies* (Washington: Government Printing Office, 1932), p. 35. See also p. 32 for Senator Dawes own statements concerning his part.
5. I have reviewed this subject at length in *The Yellowstone Story* (Boulder: Colorado Associated University Press, 1977), I, pp. 166–173.
6. "Col. J. D. Chesnut of the U.S. Army," according to Roberta C. Cheney, *Names on the Face of Montana* (Missoula: University of Montana, 1971), p. 42. His name was invariably misspelled.
7. "The Mineral Springs of the Yellowstone — Wonderful Health Restoring Qualities," *The Rocky Mountain Weekly Gazette* (Helena, MT), July 24, 1871, p.1.
8. George L. Henderson, *Yellowstone Park Manual and Guide*, July 1, 1885, p. 1.
9. George L. Henderson, *Yellowstone Park Manual*, 2nd ed. (1888), p. 2.
10. Gardner S. Turrill, *A Tale of the Yellowstone; or, In a Wagon Through Western Wyoming and Wonderland* (Jefferson, IA: G. S. Turrill Publishing Co., 1901), p. 73.
11. Alma White, *With God in the Yellowstone* (Zarepeth, New Jersey: Pillar of Fire, 1920), p. 34.
12. Henderson produced two editions, both probably printed on the *Enterprise* press at Livingston, Montana. The first was dated "July 1, 1885," and the other simply "1888."
13. Louise M. Henely, *Letters From the Yellowstone* (Grinell, IA: Ray and Frisbie, Printers, 1914), p. 6.
14. Notes copied from the personal files of Jack E. Haynes, Oct. 18, 1976.
15. Ferdinand V. Hayden, *Preliminary Report of the United States Geological Survey of Montana and Portions of Adjacent Territories, Being a Fifth Annual Report of Progress* (Washington: Government Printing Office, 1872), p. 127. The Hayden Survey's 1878 topographic map does show a lake one-quarter that size at the head of what seems to be a badly located fork of Summit Creek.
16. Trusten Peery, "Lost Lake," *Nature Notes*, XIV/9–10 (Sept.–Oct. 1937), p. 41.
17. Lee H. Whittlesey, *Yellowstone Place Names* (Helena: Montana Historical Society Press, 1988), p. 91.
18. William C. Gregg, "The Cascade Corner of Yellowstone Park," *The Outlook*, 129 (Nov. 23, 1921), p. 476.
19. G. L. Henderson, *Yellowstone National Park: Past, Present, and Future* (Washington: Gibson Bros., Printers and Bookbinders, 1891), 16 pp.
20. Ibid., p. 7. Since uric acid is an organic compound, a rerun of that analysis is in order.
21. Ibid., pp. 7–8.
22. Ibid., p. 8.

# 11

## FROM AROUND AND ABOUT



Yellowstone National Park is not an isolated entity; it is related in many ways to the surrounding communities and lines of communication. The Park's visitation has always been dependent upon transportation facilities, those outside the boundaries as well as within; and upon the development of settlements capable of providing goods and services required by visitors prior to entry as well as while sightseeing within the area. A state of interdependence has existed between the Park and certain routes and towns — of a temporary nature in some instances, but persistent in others — so that the scope of this work has to be broad enough to include the relevant place names outside the park boundaries.

In the beginning, there were two outfitting towns, or "jumping-off places," beyond which the visitor was largely on his own. These were Virginia City and Bozeman, both in Montana Territory. There was no public transportation on either route at first. The visitor who did not have his own horse, or team and wagon, could usually hire whatever was needed at those towns. Beginning in July 1874 there was service of a sort between Bozeman and Mammoth Hot Springs by way of "Zack Root's express," which left every Monday with freight and passengers for the Springs, and a short-lived stage service between Virginia City and the Lower Geyser Basin was established in the fall of 1880 by Marshall and Goff.

Those who went to the Park by the northern route found a reasonably good road, which included a section of toll road through the "Second Canyon." That road was opened by the Henderson brothers, who soon lost interest when they failed to get a franchise for a road all the way to Yellowstone Lake. The difficult, and necessary, road they had built in the "Second Canyon" passed into the hands of James George ("Yankee Jim"), whose name became attached to the canyon. This route was shorter — seventy-four miles compared to ninety-three by the "Virginia City and National Park Free Wagon Road" — and it had the advantage of terminating at a hostel of a rude kind: McCartney's cabin at Mammoth Hot Springs. Of course, the visitor was still a long way from the Lower Geyser Basin — ninety miles by a very rough trail, and fifty miles by the road Norris built in 1878. The other route, from the west, was disadvantageous in that it did not terminate at a hotel until 1880.

The construction of the Northern Pacific Railroad into Montana Territory in 1881 changed the situation rapidly. In 1882 the new railroad town of Livingston replaced the older settlement of Clark City where the Yellowstone River was crossed, and it became an important junction town with building of the Park Branch Line up the Yellowstone Valley the following year. The terminus of this branch was at another railroad town — Cinnabar, — for twenty years (until extension of the line to Gardiner in 1903).

Sporadic construction of the Mormon-controlled Utah & Northern Railway toward Montana Territory began to affect transportation on the Park's western approach after the line became part of the Union Pacific system in 1877, but the change was slower. This "Oregon Short Line" by-passed Virginia City, as it built into Montana, ultimately destroying that town's usefulness as an outfitting place for Park visitors (other towns on the line — Beaver Canyon, Monida, and finally, West Yellowstone — getting that business).

A concomitant result of construction of the Northern Pacific's branch line was the appearance of several mining towns with close ties to the Park, though of no particular importance from the viewpoint of its tourist business. They were the coal-mining towns of Horr, Aldridge, and Electric (all gone now), and the gold-mining towns of Jardine and Cooke City.

Later railroad construction east of the Park led to the appearance of another "gateway" town — Cody, Wyoming. A turn-of-the-century townsite received a railroad connection in 1901 when the Chicago, Burlington and Quincy completed its track, and the building of a road over Sylvan Pass and down the Shoshone River in 1903 made a tourist business possible at a later date. The Milwaukee system

(Chicago, Milwaukee, St. Paul and Pacific Railroad) got into the act later still by busing passengers to the West Entrance from "Gallatin Gateway" near Bozeman, Montana.

The area south of Yellowstone Park — Jackson Hole — remained roadless and cut off from Wyoming proper until the Washakie Military Road was built in 1892, across Union Pass from Fort Washakie on Wind River. This rude track was not tied into the Park road system until 1895, when the segment from South Entrance to West Thumb was completed. There was no town of importance in Jackson Hole until after World War I, and no connection to major transportation facilities until after World War II, when an airport was built.

All the railroads lost interest in the Yellowstone Park tourist business after World War II, when it became evident that the visitation was assuming an entirely different form from the prewar years. An attempt to substitute air service for the railroad train as a prime carrier of Yellowstone Park visitors has not been particularly successful, and visitation has turned ever more toward the use of private automobiles.

## BOZEMAN

This agricultural town in the eastern extremity of Montana's fertile Gallatin Valley — seventy-four miles north of Mammoth Hot Springs by the Trail Creek route — was named for John M. Bozeman, a native of Georgia, who came to the Territory as a miner in 1862. Failing to prosper in the mines, he turned to promotion of a route of travel from the Oregon Trail, across Wyoming to the Montana gold fields. While so occupied in 1863, he took note of the rich soil and abundance of water where his route debouched into the Gallatin Valley, and upon his return from an unsuccessful attempt to pilot a train of emigrants to the diggings that summer, he arranged with two friends — Daniel E. Rouse and William J. Beall — to lay out a townsite, "right in the gate of the mountains ready to swallow up all the tenderfeet that would reach the territory from the east, with their golden fleeces."[1]

The town was established in August, 1864, and a party of prospectors returning to Virginia City from the Yellowstone region that fall found six log cabins on a branch of the East Gallatin River (present Bozeman Creek). The surrounding area proved so fertile that, by the end of 1867, Bozeman had three flouring mills and the second public school established in Montana Territory. It had also replaced Gallatin City as the county seat of Gallatin County, one of the nine original counties of the territory. John Bozeman's death in April of that year at the hands of Indians east of present Livingston, Montana, created an alarm that led to the establishment of Fort Ellis (three miles east of Bozeman), with the result that the town and its adjacent farmland was safe from Indian depredation thereafter. Postal service was established at Bozeman in 1868.[2]

Bozeman, Montana. Main Street as it appeared in 1877. Photographed by Henry Bird Calfee, No. 162.

The three expeditions that, taken together, accomplished the definitive exploration of the area which is now the Yellowstone National Park — the Folsom party (1869), the Washburn party (1870), and the Hayden Survey (1871) — used Bozeman as their "jump-off" point. It was, as Folsom noted, "the last place where supplies can be obtained,"[3] and remained so until 1883, when building of the Northern Pacific Railroad's "Park Branch Line" effectively ended Bozeman's usefulness as a staging point for Park visits.

Lord Dunraven has described Bozeman's guardian fort — the place whence came the supplies and military escort for most of the official parties of exploration — as it was in the summer of 1874:

> At the upper or south end of the valley stands the clean, all-alive, and wide-awake town of Bozeman; and three miles further on, almost in the jaws of Bozeman Pass, is Fort Ellis [which] consists of a large square, two sides of which are occupied by the soldier's quarters, while the remaining side is devoted to the officer's

> houses. All along the inside of the square runs a wooden sidewalk, beside which a few unhappy trees are striving to grow; and the interior space, the centre of which is adorned with a tall flagstaff, is gravelled forming a commodious parade-ground; while the angles are flanked and protected by quaint old-fashioned looking block houses, octagonal in shape, loop-holed, and begirt with a broad balcony, upon which sentries pace everlastingly up and down. Beyond the buildings forming the square are other soldier's quarters, washerwoman's houses, stables, stores, billiard room, blacksmith's and saddler's shops and the like, the whole being surrounded by a sort of stockade fence; and furtherest removed, on a breezy elevation, are the hospital buildings, and some large stores and magazines.[4]

## VIRGINIA

There was another outfitting town serving Park visitors in the pre-railroad period — Virginia City, a mining town ninety-three miles and two mountain ranges northwest of the Upper Geyser Basin (by a very rough wagon road). This town served as a "jump-off" point for visitors coming by the southern route — that is, from the Union Pacific Railroad by way of the Montana Stage Line. For a time Virginia City competed vigorously with Bozeman for the Park tourist business, but it eventually lost out and became unimportant as access to the Park improved from the north.

The discovery of gold on July 28, 1862, on Grasshopper Creek, a headwater of Jefferson River in what was then Idaho Territory, spawned a town soon known as Bannack. The large population drawn to that area made it possible to organize a new territory — Montana — in 1864. But not all the prospectors were lucky at that first strike; many fanned out from Bannack to discoveries at other places, among them one in Alder Gulch May 26, 1863, where another town appeared.

Some Southern sympathizers, and there were many in the mines at that time, named the new town Varina, as a compliment to the charming wife of President Jefferson Davis of the Confederacy. But the local judge, G. C. Bissell, who was a Union man, would not handle papers under that name; he solved the problem by a preemptive change to Virginia City. Since there already was a town by that name in Nevada, the miners humorously compounded the duplication by calling another town at the lower end of the gulch "Nevada."[5]

Virginia City, Montana. View from the west in 1871. Photographed by William Henry Jackson, No. 187.

When Yellowstone National Park was established in 1872, the yield from the placer upon which Virginia City was founded was declining, and the people of the town were anxious to profit from travel to the new Park; however, Virginia City was at a decided disadvantage from lack of an established route into the geyser region. Gilman Sawtell had already built a rough road up the Madison River and over Raynolds Pass to his "ranch" at Henrys Lake, but from there the route over Targhee Pass and through the canyons of the Madison and Firehole Rivers was difficult even for mounted parties. So, the people of Virginia City proceeded to tilt the odds in their favor by extending the road to the Lower Geyser Basin.

During the summer of 1873 more than $2,000 was contributed to a committee for construction of a "Virginia City and National Park Free Wagon Road" (a play upon the fact that the wagon road from Bozeman to Mammoth Hot Springs included a section of toll road through Yankee Jim Canyon). Provisions, tools, and laborers were forwarded to Gilman Sawtell, and with that help he opened a rough road into the Lower Geyser Basin that fall.

Though considered a "credit to the energy of our people," the road did not do for Virginia City what its promoters had hoped. Even with extensive rebuilding by Superintendent Norris in order to facilitate mail and stage service to the Park, the Virginia City road was a failure. The town was in the wrong place, and the advance of the Utah & Northern Railway into Montana created more convenient termini for the western connection to the Park (Beaver Canyon, now Spencer, Idaho, in 1878, and Red Rock Station in Montana in 1879). The connection to the Park from the west stabilized on the road from Monida, just north of the Montana line, through Centennial Valley to Henrys Lake and a junction with the earlier road — which cut Virginia City out entirely.

Again, it is Lord Dunraven who has best described the "city" that wanted so much to be the point of departure for the new national park:

> Virginia City. Good Lord! What a name for the place! We had looked forward to a haven of rest . . . to forget our woes and weariness; an Elysium where we might be washed, clean-shirted, rubbed, shampooed, barbered, curled, cooled, and cocktailed. Not a bit of it! . . . there was no luxury. A street of straggling shanties, a bank, a blacksmith's shop, a few drygoods stores, and barrooms, constitute the main attractions of the "city." . . . The whole place was a delusion and a snare.[6]

## LIVINGSTON

The coming of the railways to Montana Territory changed the pattern of access to Yellowstone Park from both the west and the north. The Northern Pacific Railroad's tracks were laid to the crossing of Yellowstone River in 1882, and a new town — Livingston — was laid out on railroad property at the edge of Clark City.[7]

There is considerable confusion regarding the name of the railroad's town, and it is variously given as honoring different persons — each reputed to be a director of the Northern Pacific Railroad. Cheney (1971), p. 140, says the official was Crawford Livingston, of St. Paul, Minnesota (but that was obviously taken from an interview with Joe George in 1944).[8] Campbell states that the town "was named in honor of Charles Livingston of New York."[9] An official publication of the railroad indicates it was neither of those, but Johnston Livingston who was the "original director" of the Northern Pacific.[10]

By the end of November 1882, Livingston consisted of 130 houses and tents, of which ten or twelve were considered to be quite respectable. Among the total there were six hotels and restaurants, four stores, five feed stables, two butcher shops, two wholesale liquor houses, thirty saloons, and a lumber yard; the population of the place was estimated to be six hundred. The town received its post office in December of that year.

Construction of the "Park Branch Line" in 1883 established Livingston's importance to the Park tourist business, and also confirmed it as a market town for the year-around residents — an advantage Livingston still holds even though the railroad lost interest in its dwindling Park business after 1948.

Rudyard Kipling has left us a poignant description of the town of Livingston as he saw it. In the heyday of the "carriage trade," it was

> a town of 2,000 people and a junction for the little sideline that takes you to the Yellowstone National Park. It lies in a fold of the prairie and behind it is the Yellowstone River and the Gate of the Mountains through which the river flows. There is one street in the town where the cowboy's pony, and the little foal of the brood mare in the buggy, rest contentedly in the blinding sunshine while the cowboy gets himself shaved at the only barber shop and swaps lies at the bar.[11]

## CINNABAR

At the end of the Park Branch Line, fifty-one miles from the Livingston depot, was a railroad terminus that served Yellowstone Park from 1883 to 1903. It was a railhead with a strange origin and an even more peculiar ending.

The Northern Pacific Railroad had planned to build into the town of Gardiner, but a local character, Robert E. Cutler, also known as "Buckskin Jim," would not relinquish the mining claim that blocked the right-of-way immediately below town. So, the railroad dropped back three miles and built a depot and turning wye on land they were able to purchase from Joe Keeney and George Huston.[12] Tourists — persons who had purchased tickets entitling them to a package tour of Yellowstone Park including transportation, meals, and lodging — were hauled by stagecoach directly from the cars at Cinnabar to the National Hotel at Mammoth Hot Springs. Thus, the development at the railhead was minimal, as it was intended to serve the needs of freighters and local people more than visitors to the Park. The need for such a terminus was eliminated in 1903 when adjustment of property difficulties allowed extension of the Northern Pacific's rails to Gardiner.

President Roosevelt's train at Cinnabar. For a few days in 1903 the town of Cinnabar, Montana, was the capital of the United States; now it is a field with a few cellar holes in it! YNP photograph 71-18-3.

But the town of Cinnabar enjoyed one last blaze of glory before its light went out. President Theodore Roosevelt arrived at Cinnabar on April 8, 1903, and the presidential train was parked on a siding there, since the newly laid track to Gardiner was not yet ready for use. During the president's stay in the Park, a sixteen-day vacation that ended on April 24 with the dedication of the great basaltic arch opposite the rustic depot at Gardiner, Roosevelt's rolling white house (the railway car *Elysian*) remained at Cinnabar with Secretary of State William Loeb, Jr., handling presidential business from there.

The president's departure ended Cinnabar's brief period as the capital of the United States, and the few buildings too valuable to be demolished were loaded on flat cars and hauled to Gardiner. The Cinnabar townsite was incorporated into Yellowstone National Park as a part of the "Gardiner Game Ranch" addition to the Park, October 20, 1932.[13] All that remains of the town is the scar of the railroad's turning wye, and the shallow depressions of former cellars.

## RIVERSIDE/YELLOWSTONE/ WEST YELLOWSTONE

The extension of the Union Pacific System's rails from Idaho to the West Entrance of the Park in 1907 created a need for a town at the end of track on the boundary line, and the town had to have a name — which was not so easy, as it turned out.

At first the terminus was referred to merely as "the boundary," or unofficially, for the purpose of forwarding mail to the end of track, as "Yellowstone, Montana." But, upon establishment of a post office under Peter Arnet on October 23, 1908, the name became officially "Riverside, Montana." That name was unsatisfactory for several reasons: The town was not on a river, there was a well-established soldier station by that name four miles to the east, and the railroad management wanted a name identified with the Park. As a result, it was changed to "Yellowstone, Montana" on November 17, 1909, but that was even less satisfactory, leading to confusion with Park headquarters (where the post office name had been changed to Yellowstone Park, Wyoming, in 1902). The problem of misdirected mail was finally solved January 7, 1920, by yet another change — to West Yellowstone, Montana, which remains the town's name.[14]

The original settler was Joe Claus, who built a cabin in what is now the townsite during the winter of 1906–1907, probably in the hope of profiting from construction of the railway. But the nucleus from which a settlement sprang was the group of one-acre store sites abutting the railroad right-of-way that the Madison National Forest leased to Charles A. Arnet (1907), and to Samuel P. Eagle and L. A. Murray (1908).

In the latter year a six-block area, including those commercial leases, was surveyed by the Forest Service, which retained control of the land. Senator Thomas J. Walsh of Montana introduced a bill in Congress (1913) to set aside that land as a town site, and in 1919 more liberal legislation established a town of twenty-six blocks. However, no lots were sold until 1924, when squatters were allowed to buy the occupied lots for $59 each. The others were auctioned that year and in 1928.

A road was cut through the forest in 1907, to the Madison River two miles distant, where the Wylie Permanent Camping Company and the Monida and Yellowstone Stage Company built barns and other facilities required by their stages engaged in transporting the tourists into the Park from the Union Pacific depot, after scheduled service began June 1, 1909. The "Riverside Barns" were converted to a facilty for the yellow buses after the motorization of the Park in 1917, and they were razed following the Union Pacific System's abandonment of its service to the Western Entrance after 1957.

Gallatin County built a road through the northwest corner of Yellowstone Park in 1911, thus putting the new gateway town on an important north-south route that fed automobile traffic to the Western Entrance from both Bozeman, Montana, and Idaho Falls or Pocatello, Idaho. By the 1920s, West Yellowstone had become in all but a few seasons the most popular place for automobilists to enter the Park. It is an advantage the town has retained over the years, compensating very well for the loss of railroad tourism.

## CODY

A small Wyoming settlement forty-six map miles east of the Yellowstone Park boundary line owes its preeminent position as the Park's eastern gateway town to a change in the plans of a railroad.

Construction of the Chicago, Burlington and Quincy line, which was originally intended to be built to a terminus close to the East Entrance of Yellowstone Park, was halted in September, 1901, across the Shoshone River from a six-year-old Wyoming town named after one of its developers — Col. William Frederick Cody (1846–1917]) — by then better known as "Buffalo Bill." The railroad's management had second thoughts about pushing a track through the awesome gorge by which the river exited the mountains, and never advanced up the the North Fork.

The original settlement was located at some hot springs and sulphur deposits — the "Colter's Hell" of the trappers — closer to the gorge than the present town, and it became for postal purposes Cody, Wyoming, on March 24, 1896. That name was retained when the town was relocated in 1901 two and one-half miles to the east on a better site opposite the railroad terminus. Buffalo Bill also began construction of an $80,000 hotel at the new site, naming it after his sixteen-year-old daughter, Irma Louise. As opened on November 1, 1902, the Irma Hotel had "20 rooms, an elegant bar and billiard room, telegraph facilities, a telephone in every room, a barber shop, baths, steam heat, gas lights, and modern indoor plumbing. In the rear, a livery stable was equipped with all kinds of rigs, saddle and driving horses, camp and pack outfits, and offered tourist, hunting and fishing guides."[15]

Originally, the Irma was intended as the key unit of three hostels by which Buffalo Bill forwarded his important guests up the North

Fork of Shoshone River to hunting camps on its headwaters. The upper unit — Pahaska Tepee, near the East Entrance to the Park — soon became more important to the Yellowstone visitation than to Buffalo Bill's dudes (see the next section — Pahaska).

Park crews began seriously to construct a road to the East Entrance in 1901, under the direction of Engineer Hiram M. Chittenden, and the work had been advanced to the point where he could report it "opened to travel July 10, 1903," with 310 visitors entering that season. But he added: "The opinion of the officer in charge of the work is that this road will be mainly used by camping parties from the Bighorn Basin, and that therefore a good singletrack wagon road will answer every requirement until the railroad is considerably nearer the park."[16]

But the automobile changed that. By 1911 when Tex Holm began transporting guests to Pahaska in a Stanley Steamer (using the new road completed through the Shoshone gorge the previous year), the number of entries had increased to 1,524. The following year, the Holm Transportation Company of Cody was allowed to transport Burlington Route tourists into and around the Park, bringing in 209 that season — in horse-drawn vehicles, of course.

Even so, traffic on the East Entrance road was less than satisfactory from the viewpoint of the transportation concessioners. The Wylie Permanent Camping Company got into the business on that side of the Park in 1912, but did no better than the Holm Transportation Company prior to motorization of transportation in 1915. In fact, the number of entries in that year — 1,550 — was only slightly larger than in 1911.

But with the change to motorized transportation, and improvement of the road, the eastern approach gained rapidly in popularity, so that by the mid-1920s, the East and West Entrances were competitors for first place among the Park's five gates. That remained so even after the Burlington System abandoned passenger service to Cody and closed its famous Burlington Inn in 1956. Despite that loss, the town has continued to prosper — in a considerable degree because of its relationship with Yellowstone Park.

## PAHASKA

A hunting lodge on the North Fork of Shoshone River, 1.8 miles east of Yellowstone's boundary line, has always been an important way station on the eastern route into the Park.

The lodge was begun in 1901 as the upper of two overnight facilities on the rough road by which "Buffalo Bill" Cody brought guests to hunt the upper reaches of the North Fork (the other was Wapiti Inn, midway between Cody and Pahaska on that early road that passed around the south side of Cedar Mountain). At first called "Pahaska Tepee," which is the Crow Indian way of saying "Long Hair's Lodge," from the long, flowing locks Buffalo Bill wore during his Wild West Show days, the original two-story lodge building has been described this way:

> On the east and south is a large veranda-like covered porch, once lined with chairs and benches for the relaxation of visitors. Just inside was the registration desk and a small but comfortable open-beam lobby set off by a large native stone fireplace. Adding to the rustic charm, the surrounding walls were once graced with photographs, pieces of artwork, various Plains Indian regalia, handwoven Indian rugs, and numerous trophy big game heads, hides, and antlers. All of the rustic furniture was made from native materials.[17]

Eleven bedrooms, enough for the hunting clientele Buffalo Bill intended Pahaska Tepee to accommodate, occupied the upper floor, but his dream of Pahaska as the hub of a business founded on sport hunting by gentlemen of means faded quickly. However, automobilists using the eastern approach road to Yellowstone Park have insured the prosperity of this last way-stop short of the Park. The facilities have been increased over the years as the traffic required.

## MARYSVALLE/GROSVENTRE/JACKSON

The present-day metropolis of Jackson Hole had as much trouble finding a satisfactory name as West Yellowstone, but the problem was not related to the Park; in fact, the town had no direct access to Yellowstone until the nineteenth century was nearly over.

The first post office in "Jackson's Hole," the great mountain-girt valley named for fur trapper David E. Jackson, was Marysvalle, established March 25, 1892. There are local residents who believe that this original settlement was not made where the town of Jackson now stands, but at a point a few miles to the north. Postal records, however, do not reflect that; merely that the name of the post office was changed to Grosventre January 18, 1893.[18] The name was changed again three years later, on January 22, to the present usage — Jackson, Wyoming.

Originally, vehicular access to Jackson Hole was difficult. The first twelve wagons were brought in over Teton Pass in 1889 by skidding them down the precipitous east side with wheels rough-locked and treetops dragging behind. A rough freight road opened from Marysville, a former settlement a few miles east of present Ashton, Idaho, in 1894, by way of the Fall River Basin and Grassy Lake provided the first practical route into that isolated area. There was no road into the Park proper until 1896, when work on the road from West Thumb down Snake River to the south boundary of the Yellowstone Park Timberland Reserve (established by presidential order March 30 and September 10, 1891, and patrolled by soldiers of the Fort Yellowstone garrison) finally made it passable for wagons.

But it was construction of the Fort Washakie Military Road in 1899, from the Wind River Valley across Togwotee Pass to the mouth of Buffalo Fork — and a junction with Jackson Hole roads — that first opened the Park's Southern Entrance to other than local people. Visitation by way of Yellowstone's south gate remained slight until after the creation of Grand Teton National Park February 26, 1929, and it is only during the post–World War II years that Jackson has prospered as a gateway town for the sister parks to its north.

## COOKE

The last of Yellowstone's gateway towns, in respect to usefulness to its visitors, is the early mining town of Cooke City, named for Jay Cooke, financier for the Northern Pacific Railroad — in the vain hope that he would provide a branch line for hauling out the gold-rich ores of its mines.

This town, which began as "Miner's camp" in 1872,[19] tried to be "Clarks Fork City," "Galena City," and even toyed with the flowery name of "Edelweiss" before succumbing to those commercial instincts that left it with its present postal designation of Cooke City on February 1, 1882. For forty years the interest was mining — and little else — in that town of log cabins and rundown frame buildings straggling along a dirt track dominated by the high false front of the Cosmopolitan Hotel.

An idea of the 1920s for a road across the high (12,000-foot) and scenic Beartooth Plateau to connect Cooke City with Red Lodge, Montana, rescued the town from its Rip Van Winkle existence, and thus made a fifth approach to the Park a reality. This "Red Lodge Highway" was far enough along by the summer of 1933 that a visitor

Cooke City, Montana. This view of Cooke City in the 1880s shows the town when mining was the whole reason for its existence. YNP photograph, Ac. 8221.

in a 1927-model touring car was able to negotiate it — though two hours were required to go eight miles. Two years later he found the road so good he was led to speak wonderingly of the change.[20]

By the time visitors returned to Yellowstone Park following the hiatus of the war years, the metamorphosis from mining town to Western tourist orientation was so nearly complete at Cooke City that evidences of its origin were hard to find.

## MISCELLANY

The foregoing vignettes do not include two gateway towns: one — Gardiner — because it was given some notice in Chapter 2 (where the origin of the name and the town's beginnings are discussed), and the other — Gallatin Gateway — because it never amounted to much. But a little more should be said of the one, and at least something about the other.

Before the turn of the century, Gardiner was of little importance to the Park's tourist business because it was not the railhead for the Northern Pacific's Park Branch Line. A cantankerous local resident by the name of Robert E. Cutler, or "Buckskin Jim," had staked a mineral claim across the right-of-way below town, effectively barring the railroad from establishing a terminus there. As a result, the terminus remained at Cinnabar — 3.2 miles northwest of Gardiner — for twenty years, with the freight and passengers destined for the Park bypassing the town. Of course, such stubbornness was unpopular with the townspeople, and one old-timer was said to have "carried a gun" for Cutler; but they managed never to meet.

During those years, Gardiner suffered two fires (in 1886 and 1889), which probably helped its appearance — but not its reputation. A lady who passed through the town just after the second fire says: "Gardiner City; the city consisting of a dozen or so wooden shanties and log houses, chiefly burnt down, as they had a fire the night before, and the belongings of the inhabitants were strewed about all over the road. We nearly ran over a billiard table."[21]

As for the town's reputation, one Fort Yellowstone officer considered the place no more than a "collection of groggeries" that had crowded so close to the Park that the boardwalk was on the boundary line and the main street entirely within the preserve (the street still is). A soldier could get roaring drunk in Gardiner, gamble, or patronize the "house" that was reputed to have figured in Calamity Jane's old-age assistance plan. In those days it was a rough place, existing mainly for the rest and relaxation of soldiers and miners.

The extension of the Northern Pacific's track to Gardiner in 1903 did not result in the same degree of tourist orientation as occurred at West Yellowstone. Instead, the development of visitor services — the "touristy" aspect — has been balanced by concentration of concessioner warehousing and repair facilities there, and by develpment of small businesses that also served the year-round population at Park Headquarters five miles away. Also, Gardiner became a bedroom community for some Park employees.

The other "gateway" that requires some explanation was not really a gateway town. In 1926 the Milwaukee Railroad was anxious to share in the lucrative Park tour business, and having concluded an agreement with the other three railways serving the Park, it built a spur line up the West Gallatin River to a little community then known as Salesville, where the fine Gallatin Gateway Inn was built. From there, the railroad's tourists were taken by bus to West Yellowstone and into the Park. The little community of Salesville changed its name to Gallatin Gateway, and a huge arch bearing that legend spanned the gravel road leading into the Gallatin Canyon. But the Milwaukee lost interest in Park tourism after World War II; the spur line has been torn up,

Gallatin Gateway Inn. A relic of the last railroad venture into Yellowstone Park's tourist business.

the inn has passed into other hands and other uses, and the arch that once spanned Highway 191 is gone — only the town name of Gallatin Gateway remains to mark that last railroad venture into Park tourism.

As mentioned, the gold-mining town of Cooke City, just beyond the northeast corner of Yellowstone Park, was able to convert to a gateway town, but several other mining towns near the Park outlived their usefulness and have disappeared — or almost. The oldest was the coal-mining town of Horr, established in 1889 by that same Harry R. Horr who earlier had laid claim to Mammoth Hot Springs along with James C. McCartney (and also gave the springs their present name). True to form, Harry did not stay with the town that bore his name but sold his interest to the Electric Coke and Coal Company.

The new owners wanted the town identified with their concern (which took its name from the great peak towering above them). So, on January 30, 1904, the post office was renamed Electric, Montana. Local wags had their own version of *why* the name was changed. They said it was because the ladies had become tired of living in "Horr houses." Electric was a short-lived settlement; its post office closed April 30, 1915, and all that remains now to mark its site is the long line of ruined coke ovens just outside the Park boundary on the west side of Yellowstone River.

A satellite of Horr/Electric was the town of Aldridge on the shoulder of the mountain above Reese Creek. It was named for mining engineer W. H. Aldridge, and had its own post office from 1900 to 1910. Concerning its short life, Doris Whithorn says:

> Aldridge was born in 1896 on a blanket of coal . . . in time there arose a post office, a school, several stores, boarding houses, and saloons. Within months 800 people were living in Aldridge. As it came into being it stretched and yawned, then shriveled and died. Fifteen years was its lifetime in the moments of eternity.[22]

Aldridge was of interest in Yellowstone Park affairs mainly for contributing more than its fair share of poachers and stagecoach robbers to annoy the authorities.

## NOTES

1. Merrill G. Burlingame, *The Montana Frontier* (Bozeman: Big Sky Books, Montana State University, 1980), p. 342.
2. Roberta Carkeek Cheney, *Names on the Face of Montana* (Missoula: University of Montana, 1971), p. 25.
3. Charles W. Cook et al., ed. by Aubrey L. Haines, *The Valley of the Upper Yellowstone* (Norman: University of Oklahoma Press, 1965), p. 11.
4. Windham T.W.Q. Dunraven, *The Great Divide* (London: Chatto and Windus, 1876), pp. 55–56.
5. Burlingame, op. cit., p. 87; see also, Cheney, op. cit., p. 234.
6. Dunraven, op. cit., p. 43.
7. The use of "city" in the naming of frontier towns seldom had any more significance than an expression of optimism. In this case, speculators had seized upon ground where the survey of the railroad's main line crossed the Yellowstone in the hope that it would prove to be a valuable town site. There had been an earlier settlement — Bensons Landing, named for Charles Benson — about three miles downriver from the railroad crossing.
8. Interviewed by J. E. Haynes at Bozeman, Montana, February 12, 1944. Joe George, 90 years old at the time of the interview, was the son of John P. George, the original settler on the Crow Indian Reservation.
9. Marius R. Campbell, *Guidebook of the Western United States; Part A, The Northern Pacific Route, With a Side Trip to Yellowstone Park* (Washington: Government Printing Office, 1915), p. 88.
10. E. V. Smalley, Compiler, *Northern Pacific Railroad; Book of Reference, for Use of the Directors and Officers of the Company* (New York: E. Wells, Sackett & Rankin, 1883), p. 435.
11. Rudyard Kipling, *From Sea to Sea* (Garden City, NY: Doubleday, Page & Co., 1920), p. 62.

12. Aubrey L. Haines, *The Yellowstone Story* (Boulder: Colorado Associated University Press, 1977), I, p. 267.
13. Ibid., II, pp. 331–332.
14. Sam and Ed Eagle, *West Yellowstone's 70th Anniversary, 1908 to 1978* (West Yellowstone, MT: Eagle Company, Inc., 1978), Part 2, p. 6. Jack E. Haynes, *Haynes Official Guide, Yellowstone National Park*, (St. Paul, MN: Published & Illustrated by F. J. Haynes, 1915), p. 40, notes that construction stopped at the boundary November 12, 1907.
15. Bill Barnhart, *The Northfork Trail* (Cody, Wyoming: Elkhorn Publishing, 1982), p. 4.
16. Major H. M. Chittenden, *Report of the Acting Superintendent of the Yellowstone National Park to the Secretary of the Interior* (Washington: Government Printing Office, 1904), p. 14.
17. Barnhart, op. cit., p. 42.
18. Pronounced "GROW-vaunt," the name identified a subgroup of the Blackfeet Confederacy and is best translated from Canadian French as "Big Bellies."
19. William Blackmore, "Diary — 1872 — Fourth Visit to the United States," (transcription by Brayer, Western Range Cattle Industry Study, State Museum, Denver, 1945), No. 6, pp. 82, 85 & 89.
20. Emil Kopac to Chief Park Naturalist David deL. Condon, August 15, 1949. Correspondence received, Yellowstone National Park. The serious advocacy of this approach began with an item in the *Livingston* (MT) *Enterprise*, April 14, 1920. See "Plan Entrances for Cooke Road."
21. Georgina M. Synge, *A Ride Through Wonderland* (London: Low, Marston & Co., 1892), pp. 131–132.
22. Quoted in Cheney, *Names* (1971), p. 3; see Note 2 for citation.

# SOURCES

Adams, Robert, Jr. "The Tetons Scaled." *Philadelphia Inquirer,* September 2, 1872.

Albright, Horace M. "Annual Report for Yellowstone National Park, 1928." Mimeographed, Yellowstone Park Reference Library.

Allen, Charles J. "Map of the Yellowstone National Park, Showing Roads Completed, Under Construction, and Approximately Projected, November 14, 1888." Original tracing, Yellowstone Park Reference Library.

Allen, E. T., and A. L. Day. *Hot Springs of the Yellowstone National Park.* Carnegie Institution Publication No. 466. Washington, 1935.

Anderson, George S. *Report of the Acting Superintendent of the Yellowstone National Park to the Secretary of the Interior, 1894.* Washington, 1894.

——— *Report of the Acting Superintendent of the Yellowstone National Park to the Secretary of the Interior, 1896.* Washington, 1896.

Bandy, William R. "Plat of Survey of the East Boundary of Yellowstone National Park," September 27, 1931.

Barlow, John W. "Natural Wonders." *Chicago Evening Journal,* January 13, 1872.

Barlow, John W., and David P. Heap. *Report of a Reconnoissance of the Upper Yellowstone in 1871.* Senate Executive Documents, 42d Cong., 2n Sess., No. 66, (1872).

Barnhart, Bill. *The Northfork Trail.* Cody, WY, 1982.

Batterman, C. S., to Capt. G. S. Anderson. August 2, 1893. Doc. 681, Yellowstone National Park Archives.

Bauer, C. Max. "Place Names of Yellowstone National Park." 1937 typescript. Yellowstone Park Reference Library.

Beal, Merrill D. *The Story of Man in Yellowstone.* Yellowstone National Park, 1960.

Blackmore, William. "Diary — 1872 — Fourth Visit to the United States," Transcription by Brayer. Colorado State Museum, Denver, 1945.

Blaine, John E. "Map of the Territory of Montana to Accompany the Report of the Surveyor General . . . 1871." National Archives, RG–49, GLO(OMF)-Montana 5.

Bliss, A. A. (Tony). *Silver Tip Ranch, 1922–47.* Privately printed, 1947.

Brackett, Mrs. William Starr. Oral history transcript, November 9, 1950. Yellowstone Park Reference Library.

Brett, Lloyd M. *Report of the Acting Superintendent of the Yellowstone National Park, 1913.* Washington, 1913.

Bryan, T. Scott. *The Geysers of Yellowstone.* Boulder, Colorado, 1979.

Burdet, Bob. "In the National Park." *Enterprise.* Livingston, MT, July 28, 1888.

Burlingame, Merrill G. *The Montana Frontier.* Bozeman, MT, 1980.

Campbell, Marius R. *Guidebook of the Western United States; Part A, The Northern Pacific Route, With a Sidetrip to Yellowstone Park.* Washington, 1915.

Campbell, Reau. *Campbell's New Revised Guide and Descriptive Book of the Yellowstone.* Chicago, 1914.

Carroll, Lewis [Charles L. Dodgson]. *Alice's Adventures in Wonderland.* Oxford, U.K., 1866.

Catlin, George. *Letters and Notes on the Manners, Customs, and Conditions of the North American Indians,* 2 Vols. New York, 1841.

Cheney, Roberta C. *Names on the Face of Montana.* Missoula, MT, 1971.

Chittenden, Hiram M., to Acting Superintendent George S. Anderson. March 28, 1895. Yelowstone Archives.

———. *The Yellowstone National Park.* Cincinnati, 1895.

———. "Construction, Repair, and Maintenance of Roads and Bridges in the Yellowstone National Park," *Annual Report of the Chief of Engineers for 1900.* Washington, 1900.

———. "Annual Reports Upon the Construction, Repair, and Maintenance of Roads and Bridges in the Yellowstone National Park." *Annual Report of the Chief of Engineers for 1901.* Washington, 1901.

———. *The American Fur Trade of the Far West,* 2 Vols. New York, 1902.

———. "Map of the Grounds in the Vicinity of Mammoth Hot Springs, in the Yellowstone National Park . . . 1903." *Report of the Acting Superintendent of the Yellowstone National Park to the Secretary of the Interior, 1903.* Washington, 1903.

———. *The Yellowstone National Park.* 4th ed. Cincinnati, 1903.

———. "Improvement Yellowstone National Park," *Report of the Acting Superintendent of the Yellowstone National Park to the Secretary of the Interior, 1904.* Washington, 1904.

———. *The Yellowstone National Park.* 6th ed. Cincinnati, 1910.

———, *The Yellowstone National Park.* Ed. by Richard A. Bartlett. Norman, OK, 1964.

———. "The 'National Park Range' of Mountains; What is its Rightful Name?" Unpublished MS. [unsigned but agreeing in style and preparation with work of Chittenden]. Yellowstone Park Reference Library *(no date).*

Chittenden, Hiram M., and Alfred T. Richardson. *Life, Letters and Travels of Father Pierre Jean DeSmet, S. J.: Missionary Labors and Adventures Among the Wild Tribes of the North American Indians.* 4 Vols. New York, 1905.

*Christian Science Monitor.* "The Yellowstone Trail." June 12, 1917.

Clark, William. *A Map of Part of the Continent of North America . . . Compiled from the information of the best informed travellers through that quarter of the Globe . . . [1806–1812].* MS No. 303-iv, Coe Collection. Beinecke Library, Yale University.

———. *A Map of Lewis and Clark's Track Across the Western Portion of North America, from the Mississippi to the Pacific Ocean.* Copied by Samuel Lewis . . . 1806. National Archives, RG–77, Office of the Chief of Engineers, US 529.

———. *History of the Expedition Under the Command of Captains Lewis and Clark.* Ed. by Nicholas Biddle. Philadelphia, 1814.

Clawson, C. C. "Notes on the Way to Wonderland." *New North-West,* Deer Lodge, MT September 9, 1871, through February 24, 1872.

Cockran, J.M.V. "The Naming of the Lone Star Geyser." *Yellowstone Nature Notes.* Oct.–Nov. 1932.

Cook, Charles W. "Valley of the Upper Yellowstone." *Western Monthly.* July 1870.

Cook, Charles W., David E. Folsom, and William Peterson. *The Valley of the Upper Yellowstone.* Ed. by Aubrey L. Haines. Norman, OK, 1965.

Coolidge, [John] Calvin, to Charles W. Cook. February 14, 1924. (Original held by Mrs. Josephine Cook Mueller. Lewistown, MT).

Craighill, William E. "Report of Lieutenant William E. Craighill, Corps of Engineers." *Annual Report of the Chief Of Engineers for 1890.* Washington, 1890.

Cramton, Louis C. *Early History of Yellowstone National Park and its Relationship to National Park Policy.* Washington, 1932.

Daugherty, J. S. "Sketch of Part of the Missouri & Yellowstone Rivers with a Description of the Country," 2 Sheets. National Archives, RG–92, Rec. of Off. QM Gen.

Davis, John W. "Yellowstone Park." *Courier-Journal.* Louisville, Kentucky, April 18, 1884.

deLacy, Walter W. *Map of the Territory of Montana and Portions of Adjoining Territories . . . 1865.* St. Louis, New York, and Baltimore, 1865.

———. *Map of the Territory of Montana, With Portions of the Adjoining Territories.* New York, 1870.

———. Letter to Martin Maginnis. May 9, 1874. Montana Historical Society, Helena.

DeSmet, Pierre J. *Chart of the Head of Yellowstone.* Drawn by Jim Bridger and captioned by DeSmet (1851). Jesuit Archives, Missouri Province. IX: Desmetiana, C–8: Atlas, No. 10. St. Louis University.

———. [Untitled map, 1851]. Jesuit Archives, Missouri Province. IX: Desmetiana, C–8: Atlas, No. 5. St. Louis University.

Doane, Gustavus C. *Map of the Route of the Yellowstone Expedition Escort Commanded by Lieut. G. C. Doane, U.S.A., Sept., 1870.* National Archives, RG–77; Office of the Chief of Engineers, Q 329 — No. 30.

———. *Official Report of the Washburn-Langford-Doane Expedition into the Upper Yellowstone in 1870.* MS 492. Montana State University, Bozeman.

Doubt, Dorothea G. to "The Secretary of the Museum." September 7, 1933. MS, Yellowstone Park Reference Library.

Drake, O.S.T. "A Lady's Trip to the Yellowstone Park." *Every Girl's Annual.* London 1887.

Dudley, W. H. *The National Park From the Hurricane Deck of a Cayuse, or the Leiderkrantz Expedition to Geyserland.* Butte City, MT, 1886.

Dunraven, Windham T.W.Q. *The Great Divide: Travels in the Upper Yellowstone in the Summer of 1874.* London 1876.

Eagle, Sam and Ed. *West Yellowstone's 70th Anniversary, 1908 to 1978.* West Yellowstone, MT, 1978.

Editorial. "The Scenery of the Yosemite . . ." *Montana Post.* Virginia City, MT, July 14, 1866.

———. "The Yellowstone Mountains." *Montana Post.* Virginia City, MT, February 16, 1867.

———. "Interesting from the Yellowstone." *Daily Herald.* Helena, MT, April 11, 1867.

———. "Who Will Now Say We Are Not in Danger?" *Montana Post.* Virginia City, MT May 18, 1867.

———. "Army News." *Montana Post.* Virginia City, MT, June 29, 1867.

———. "The Recent Massacres . . ." *Semi-Weekly Mining News.* Salmon City, I.T., September 1, 1867.

———. "The Headquarters of the Yellowstone." *Weekly Herald.* Helena, MT, December 12, 1867.
———. "Good Story — We clip from an eastern paper . . ." *Weekly Herald.* Helena, MT, December 26, 1867.
———. "Expedition to the Yellowstone." *Weekly Herald.* Helena, MT, July 29, 1869.
———. "A Thrilling Event on the Yellowstone." *Daily Herald.* Helena, MT, May 18, 1870.
———. "The Yellowstone Expedition." *Daily Herald.* Helena, MT, September 26, 1870.
———. "The Yellowstone Expedition." *Daily Herald.* Helena, MT, September 27, 1870.
———. "The Yellowstone Expedition," *Daily Herald.* Helena, MT, September 28, 1870.
———. "The Great Falls of the Yellowstone." *Daily Oregonian.* Portland, OR, September 29, 1870. (Reprinted from the Bozeman *Montana Pick and Plow*).
———. "Yellowstone Papers." *Rocky Mountain Weekly Gazette.* Helena, MT, October 19, 1870.
———. "Yellowstone Papers." *Rocky Mountain Weekly Gazette.* Helena, MT, October 24, 1870.
———. "A National Park for Washington." *Daily Morning Chronicle.* Washington, January 20, 1871.
———. "The Mineral Springs of the Yellowstone . . ." *The Rocky Mountain Weekly Gazette.* Helena, MT, July 24, 1871.
———. "Hot Springs . . ." *Avant Courier.* Bozeman, MT, September 13, 1871.
———. "The New Wonderland." *Daily Times.* New York, October 23, 1871.
———. "The Hon. N. P. Langford of Montana . . ." *Territorial Enterprise.* Virginia City, NV, January 1, 1872.
———. "Our National Park." *Daily Herald.* Helena, MT, February 28, 1872.
———. [Establishment of Yellowstone Park]. *The Nation.* March 7, 1872.
———. "The Great Survey." *Philadelphia Press.* Philadelphia, September 13, 1872.
———. "Tour Through Montana." *Avant Courier.* Bozeman, MT, October 23, 1879.
———. "The Mines of Southeastern Montana." *Avant Courier.* Bozeman, MT, February 12, 1880.
———. "The Yellowstone Park — Its Management." *Avant Courier.* Bozeman, MT, September 30, 1880.
———. "Wonderland." *Avant Courier.* Bozeman, MT, January 1, 1881. (Reprinted from *Omaha Republican*).
———. "There are fifty to sixty persons . . ." *Avant Courier.* Bozeman, MT, August 4, 1881.
———. "The Sheridan Geyser." *Avant Courier.* Bozeman, MT, September 15, 1881.
———. "Newsy Letter from Upper Yellowstone." *Avant Courier.* Bozeman, MT, April 6, 1882.
———. "Gardiner." *Daily Enterprise.* Livingston, MT, June 6, 1883.
———. "Near the Stalactite Caves . . ." *Daily Enterprise.* Livingston, MT, June 14, 1883.
———. "The Yellowstone Park House." *Daily Enterprise.* Livingston, MT, June 16, 1883.

———. "G. L. Henderson, Assistant Superintendent . . ." *Daily Enterprise.* Livingston, MT, July 16, 1883.
———. "Local Layout." *Daily Enterprise.* Livingston, MT, August 20, 1883.
———. "The Park Branch . . ." *Daily Enterprise.* Livingston, MT, September 3, 1883.
———. "Work in the Park." *Daily Enterprise.* Livingston, MT, October 5, 1883.
———. "The Name Yellowstone." *Daily Enterprise.* Livingston, MT, October 23, 1883.
———. "From Mr. Marshall . . ." *Daily Enterprise.* Livingston, MT, October 30, 1883.
———. "Cooke City News." *Daily Enterprise.* Livingston, MT, November 14, 1883.
———. "Park Paragraphs." *Daily Enterprise.* Livingston, MT, November 21, 1883.
———. "Park Paragraphs." *Daily Enterprise.* Livingston, MT, November 28, 1883.
———. "The Road Through the Canyon." *Daily Enterprise.* Livingston, MT, February 9, 1884.
———. "One of the Illustrations . . ." *Daily Enterprise.* Livingston, MT, March 11, 1884.
———. "J. F. Yancey, of . . . Park." *Daily Enterprise.* Livingston, MT, March 24, 1884.
———. "A Railroad Through Clark's Fork Canyon." *Daily Enterprise.* Livingston, MT, May 27, 1884.
———. [Johnston Livingston]. *Daily Enterprise.* Livingston, MT, June 9, 1884.
———. "Park Notes." *Weekly Enterprise.* Livingston, MT, November 15, 1884.
———. "The Geysers in Winter." *Weekly Enterprise.* Livingston, MT, December 6, 1884.
———. "The Park Arrests." *Weekly Enterprise.* Livingston, MT, December 20, 1884.
———. "Park Notes." *Weekly Enterprise.* Livingston, MT, April 25, 1885.
———. "F. F. Fridley is fitting up . . ." *Weekly Enterprise.* Livingston, MT, May 30, 1885.
———. "Doings in Wonderland." *Weekly Enterprise.* Livingston, MT, June 13, 1885.
———. "Local Layout." *Weekly Enterprise.* Livingston, MT, August 1, 1885.
———. "Local Layout." *Weekly Enterprise.* Livingston, MT, August 31, 1885.
———. "Park Items." *Weekly Enterprise.* Livingston, MT, November 21, 1885.
———. "From Mammoth Hot Springs." *Weekly Enterprise.* Livingston, MT, December 12, 1885.
———. "Excelsior Geyser." *Weekly Enterprise.* Livingston, MT, June 9, 1888.
———. "Yellowstone Park." *Pioneer Press.* St. Paul, MN, June 4, 1892.
———. "Our Towns and What They Are." *Wonderland.* Gardiner, MT, May 17, 1902.
———. [LaDuke's Mineral Hot Springs]. *Wonderland.* Gardiner, MT, August 21, 1902.
———. "The Corner Stone Was Laid." *Wonderland.* Gardiner, MT, April 30, 1903.
———. "Park to Park Highway Will be Built." *Daily Enterprise.* Livingston, MT, June 29, 1911.
———. "New Lake in Wonderland." *Daily Enterprise.* Livingston, MT, August 1, 1911.
———. "Plan Entrances For Cooke Road." in *Daily Enterprise.* Livingston, MT, April 14, 1920.

———. "A Great Figure of the Yellowstone — General Hiram M. Chittenden." *Daily Enterprise.* Livingston, MT, June 30, 1923.
———. [Firing line on Deckard Flat]. *Leader.* Great Falls, MT, January 14, 1936.
———. "Lost Lake." *Yellowstone Nature Notes.* Sep.–Oct. 1937.
Ericson, Mildred. "Animal Accidents in Geyserland." *Nature Magazine.* June-July 1948.
Erwin, James B. *Report of the Acting Superintendent of the Yellowstone National Park to the Secretary of the Interior, 1898.* Washington, 1898.
Evans, John. [MS map of the Missouri River compiled in 1797]. Original is No. 303-iii, Coe Collection, Beinecke Library, Yale University.
Everts, Truman C. "Thirty-seven Days of Peril." *Scribner's Monthly.* November 1871.
Fenio, Ray. Unofficial place names identified in September, 1981. Author's file.
[Ferris, Warren A.]. "Rocky Mountain Geysers." *Western Literary Messenger.* July 13, 1842.
[———.]. "Rocky Mountain Geysers." *The Wasp.* Nauvoo, Illinois, August 13, 1842.
[———.]. "Visit To The Yellowstone Geysers 41 Years Ago." *Weekly Independent.* Helena, MT, May 1, 1874.
———. *Life in the Rocky Mountains 1830–1835.* Ed. by P. C. Phillips. Denver, 1940.
Finck, Henry T. "Yellowstone Park As a Summer Resort." *The Nation*, September 27, 1900.
Folsom, David E. *The Folsom-Cook Exploration of the Upper Yellowstone, 1869.* St. Paul, 1894.
Fort Yellowstone. "Record Book of Interments, Post Cemetery, Fort Yellowstone, Wyo." Yellowstone Archives, Doc. 155. Yellowstone Park Reference Library.
Freeman, Legh R. "California." *The Frontier Index.* Laramie City, D. T., May 5, 1868.
Gannett, Henry. *Boundaries of the United States,* 3rd ed. (U.S. Geological Survey, Bull. No. 226, Series F, Geography). Washington, 1904.
Gallatin County, MT Deeds, 1869–1872. Gallatin County Courthouse, Bozeman, MT.
General Land Office. [MS map prepared by Surveyor-General Meredith, 1869]. National Archives, RG–49, GLO(OMF)-Montana 2.
———. [Map] *State of Idaho,* 1921. Washington, 1921.
George, Joe. Interview by Jack E. Haynes, February 12, 1944. Bozeman, MT.
Gerrish, Theodore. *Life in the World's Wonderland.* Biddeford, MA, 1887.
Gibbon, John. "The Wonders of the Yellowstone." *Journal of the American Geographical Society of New York,* (1874).
Gibson, Francis M. [MS pocket diary kept on a trip through Yellowstone National Park, Aug. 7 to 27, 1882] Yellowstone Park Reference Library.
Gibson, John. *Map of North America,* 1763. St. Louis University Library.
Gillette, Warren C. *Diary of Warren Caleb Gillette from August 17, 1870 to September 27, 1870.* MS. Montana Historical Society, Helena.
Glover, Richard. *David Thompson's Narrative, 1784–1812.* Toronto, Canada, 1962.
Gondos, Victor to Jack E. Haynes, March 7, 1955. Excerpt in author's file.
Goode, George W. *Report of the Acting Superintendent of the Yellowstone National Park to the Secretary of the Interior, 1900.* Washington, 1900.

Gourley, James. [Reminiscent account of a trip to Cooke City and the Lamar River area in 1870]. Shows two dates: June 8, 1870, and March 28, 1929. Yellowstone Park Reference Library.

Greenhow, Robert. *Map of the Western & Middle Portions of North America, to Illustrate the History of California, Oregon, and Other Countries on the Northwest Coast of America.* St. Louis University, 1844 (Date added in pencil).

Gregg, William C. "The Cascade Corner of Yellowstone Park." *The Outlook.* November 23, 1921.

Gregory, James F. "Report of Lieut. Col. James F. Gregory." *Report of Lieut. General P. H. Sheridan.* September 20, 1881. Washington 1882.

Gunnison, John W. *The Mormons, or Latter-Day Saints.* Philadelphia 1852.

Guptill, A. B. *Haynes Guide to Yellowstone Park.* St. Paul, MN, 1896.

———. *Haynes Guide to Yellowstone Park.* St. Paul, MN, 1899.

———. *Haynes Guide to Yellowstone Park.* St. Paul, MN, 1903.

Hague, Arnold. *Geological History of the Yellowstone National Park.* Reprinted from *Transactions* of the American Institute of Mining Engineers (1888).

———. "Soaping Geysers," (1889). *Annual Reports of the Board of Regents of the Smithsonian Institution, July 1892.* Washington, 1893.

———. *Atlas to Accompany Monograph XXXII on the Geology of the Yellowstone National Park.* Washington, 1904.

———. *Geological History of the Yellowstone National Park.* Washington, 1928.

Haines, Aubrey L. "The Bridge That Jack Built." *Yellowstone Nature Notes,* January–February 1947.

———. "McGuirk's Medicinal Springs." *Yellowstone Nature Notes,* March–April 1947.

———. "Visit to the Queen's Laundry Bathhouse." A typed report to the Chief Park Naturalist. Yellowstone National Park, June 14, 1961.

———. "Memorandum for the Park Historian's File — Location of the Turkey Pen Cabin." Yellowstone National Park, August 28, 1962.

———. "The Park Branch Line: Doorway to the Yellowstone" (mimeographed). Yellowstone National Park, 1963.

———. "Memorandum to the Chief Park Naturalist — On the Subject of the Gardiner Game Ranch Addition." Yellowstone National Park, May 4, 1967.

———. *Yellowstone National Park: Its Exploration and Establishment.* Washington, 1974.

———. Notes copied from the personal files of Jack E. Haynes, October 18, 1976.

———. *The Yellowstone Story,* 2 vols. Boulder, Colorado, 1977.

———. Personal recollections from the period 1938–1969.

Hals, Carl J., and A. Rydstrom. *Map of the Yellowstone National Park,* 1882.

Hamilton, Mrs. James M. "Through Yellowstone Park in 1883 with Mrs. James M. Hamilton" (Typed). Interview at the meeting of the Montana State University History Group, January 1967.

Harris, Burton. *John Colter: His Years in the Rockies.* New York, 1952.

Harris, Moses. *Report of the Superintendent of the Yellowstone National Park to the Secretary of the Interior.* Washington, 1887.

Harrison, Benjamin. Presidential Executive Orders of March 30 and September 10, 1891.

Harrison, Carter. *A Summer's Outing and the Old Man's Story.* Chicago, 1891.
"Hattie." Letter written from Helena, MT, to Mrs. Mary Shulber, Dexter, Iowa, September 4, 1887 (Copy). Yellowstone Park Reference Library.
Hauser, Samuel T. [diary, August 17 to September 24, 1870]. MS No. 259, Coe Collection, Beineke Library, Yale University.
Hayden, Ferdinand V. [Map proposing a Yellowstone National Park, 1871]. National Archives, RG–57, Hayden Survey, YNP.
———. *Preliminary Report of the United States Geological Survey of Montana and Portions of Adjoining Territories, Being a Fifth Annual Report of Progress.* Washington, 1872.
———. *Sixth Annual Report of the United States Geological Survey of the Territories, 1872.* Washington, 1873.
———. *Twelfth Annual Report of the United States Geological Survey of the Territories,* 2 parts. Washington, 1883
Haynes, Frank J. *Souvenir, Yellowstone National Park.* Fargo, D. T., 1889.
Haynes, Jack E. *Haynes Official Guide, Yellowstone National Park.* St. Paul, MN, 1915.
———. *Haynes New Guide And Motorists' Complete Road Log of Yellowstone National Park.* St. Paul, MN, 1921.
———. *Haynes New Guide And Motorists' Complete Road Log of Yellowstone National Park.* St. Paul, MN, 1922.
———. *Haynes New Guide And Motorists' Complete Road Log of Yellowstone National Park.* St. Paul, MN, 1928.
———. *Haynes Guide, The Complete Handbook of Yellowstone National Park.* St. Paul, MN, 1936.
———. *Haynes Guide, Handbook of Yellowstone National Park.* St. Paul, MN, 1939.
———. *Haynes Guide, Handbook of Yellowstone National Park.* St. Paul, MN, 1940.
———. *Haynes Guide, Handbook of Yellowstone National Park.* St. Paul, MN, 1942.
———. Interview with Joe George at Bozeman, Montana, February 12, 1944. Haynes Papers, Montana State University. Bozeman.
———. *Haynes Guide, Handbook of Yellowstone National Park.* Bozeman, MT, 1947.
———. "Memorandum to the files, Jan. 8, 1953, concerning territorial post offices." Haynes Papers, Montana State University. Bozeman.
———. Letter to William R. Bandy, Billings, MT, June 2, 1953. Haynes Papers, Montana State University. Bozeman.
———. Note recorded during an interview with Joseph Joffe, May 8, 1956. Haynes Papers, Montana State University. Bozeman.
———. *Haynes Guide, Handbook of Yellowstone National Park.* Bozeman, MT, 1958.
———. *Yellowstone Stage Holdups.* Bozeman, MT, 1959.
———. *Haynes Guide, Handbook of Yellowstone National Park.* Bozeman, MT, 1961.
———. Reminiscences, 1953 to 1963. Author's file.
———. *Haynes Guide, Handbook of Yellowstone National Park.* Bozeman, MT, 1966.
Hazeltine, Harry (donor). "Mother's Journal of a Trip Thru Yellowstone & Teton." MS notebook, June 26 to October 20, 1901. Yellowstone Park Reference Library.
Heath, Charles A. *A Trial of a Trip; Notes from the Journal of Charles A. Heath.* Chicago, 1905.
Heaton, Augustus G. *Yellowstone Letters.* New York, 1906.
Hedges, Cornelius. "Mount Everts." *Daily Herald.* Helena, MT, October 8,1870.

———. "Diary of Cornelius Hedges from June 24, 1870 to October 16, 1871." MS No. 2165, Montana Historical Society, Helena.

———. "Address delivered by Cornelius Hedges on Sunday evening January 29th, 1871, eulogistic of Hon. Henry D. Washburn, at the M.E. Church, Helena, M.T." Copy in the Hedges Papers. Montana Historical Society, Helena.

Henderson, A. Bart. "Journal of the Yellowstone Expedition of 1866 Under Captain Jeff Standifer; Also the Diaries Kept by Henderson During His Prospecting Journeys in the Snake, Wind River, and Yellowstone Country During the Years 1866–72." Original transcription by Granville Stuart. MS No. 452, Coe Collection, Beinecke Library, Yale University.

Henderson, C. Hanford. "Through the Yellowstone on Foot." *Outing*, May 1899.

Henderson, George L. [MS journal kept at Mammoth Hot Springs, 1882–1886]. Yellowstone Archives, No. 141.

———. *Yellowstone Park Manual And Guide.* Mammoth Hot Springs, Yellowstone National Park, July 1, 1885.

———. "Haynes Winter Expedition." *Independent.* Helena, MT, February 6, 1887.

———. *Yellowstone Park Manual*, 2nd ed. Mammoth Hot Springs, Yellowstone National Park, 1888.

———. "Wonders in Wyoming." *Pioneer Press.* St. Paul, MN, December 25, 1888.

———. "Naming a Geyser." *Daily Herald.* Norristown, [?], January 19, 1889.

———. *Yellowstone Park: Past, Present, and Future.* Washington, 1891.

———. "Berger's Geyser." *Post.* Livingston, MT, March 8, 1900.

———. "Rapid Transit in the Park." *Post.* Livingston, MT, March 10, 1900.

———. "On the Terraces . . ." *Enterprise.* Livingston, MT, June 22, 1901.

Henely, Louise M. *Letters from the Yellowstone.* Grinell, IA, 1914.

Holmes, William H. Excerpts from his diary for the winter, spring and most of the summer of 1878. Copy in the Yellowstone Park Reference Library.

Holt, Peter. [Diary of Peter Holt, July 2, 1889 to April 27, 1900]. Scout Diaries — Yellowstone Archives.

Hood, Washington. *Map exhibiting the practicable passes of the Rocky Mountains; together with the Topographical features of the country adjacent to the headwaters of the Missouri, Yellowstone, Salmon, Lewis and Colorado Rivers . . . 1839.* National Archives, RG–77, Records of the Office of the Chief of Engineers, Headquarters Map File, U.S. 110.

Horr, Harry R. Letter to Capt. George L. Anderson, Acting Superintendent of Yellowstone National Park, May 16, 1891. Yellowstone Archives, Doc. 1050.

Hyde, John. *Official Guide to the Yellowstone National Park.* St. Paul, MN, 1886.

Irving, Washington. *The Works of Washington Irving*, 3 Vols. New York, 1882.

Jaager, T. A. "A Natural Bear Trap." *Appelton's Popular Science Monthly.* February 1889.

Jackson, William H. *Time Exposure.* New York, 1940.

Jenkins, Henry D. "The Lucky Cowboy." MS copied on magnetic tape by A. L. Haines, July 3, 1961.

Joffe, Joseph. "The Grand Old Man of Yellowstone National Park, John W. Meldrum." Typescript, 1940. Copy in Yellowstone Park Reference Library.

Johnson, A. J. *Johnson's Nebraska, Dakota, Idaho and Montana* [map]. New York, 1865.

Jones, William A. *Map of a Military Reconnoissance, in North Western Wyoming, 1873.* Washington, 1874.

———. *Report Upon the Reconnaissance of Northwestern Wyoming, Including Yellowstone National Park, Made in the Summer of 1873.* Washington, 1875.

———. *Yellowstone National Park* [map], November 30, 1892. Original in Yellowstone Park Reference Library.

———. *Road Map* [Yellowstone National Park], 1894. Original in Yellowstone Park Reference Library.

Josephy, Alvin M., Jr. *The Nez Perce Indians and the Opening of the Northwest.* New Haven, 1965.

"Journal History." Church Historian's Office, Church of Jesus Christ of Latter-Day Saints. Salt Lake City.

Joyner, Newell F. [Notes collected by Newell Joyner concerning various explorers and sites, c. 1930]. Yellowstone Park Reference Library.

Kendall, W. C. *The Fishes of the Yellowstone National Park.* Washington, 1915.

Kennerly, William C. *Persimmon Hill: A Narrative of St. Louis.* Norman, OK, 1948.

Kilmartin, J. O. Kilmartin, Board on Geographic Names, to Director, National Park Service, May 16, 1962. Copy to Yellowstone National Park.

———. Kilmartin, Board on Geographic Names, to Historian Richard A. Bartlett, May 2, 1966. Copy in author's file.

Kipling, Rudyard. *From Sea to Sea.* New York, 1920.

Knowlton, F. H. "A Gulch of Death." *Popular Science News.* Undated clipping, Scrapbook 2, Yellowstone Park Reference Library.

Kopac, Emil. Letter to Chief Park Naturalist David deL. Condon. August 15, 1949.

Langford, Nathaniel P. [Notes of a lecture given by N. P. Langford during the winter of 1870–71]. MS, Yellowstone Park Reference Library.

———. "Ascent of Mount Hayden." *Scribner's Monthly,* June 1873.

———. *Diary of the Washburn Expedition to the Yellowstone and Firehole Rivers in the Year 1870.* St. Paul, MN, 1905.

Larocque, Francois A. "A Few Observations on the Rocky Mountains Indians with Whom I Passed the Summer." *The Oregon Country Under the Union Jack,* ed. by B. C. Payette. Montreal, Canada, 1962.

Larson, John. [Monthly reports of Cpl. John Larson, Co. I, 6th Cav., in charge of detachment at Norris Station, to the Commanding Officer, Fort Yellowstone, for the months of May through December, 1895]. Yellowstone Archives.

LeHardy, Paul, Sr. "An Autobiography of Paul LeHardy." Portion covering experiences while a member of the Jones Expedition into the Yellowstone country in 1873. Copy, Yellowstone Park Reference Library.

*Leslie's Illustrated Newspaper.* "Great Soda Mountain and Jupiter's Baths, in the Yellowstone Region." September 31, 1871.

Lindsley, Chester A. [Letter to William M. Nichols, May 6, 1914]. Yellowstone Archives.

Little, Raymond G. Interviewed by A. L. Haines, April 21, 1961. Audio Tape 61–1. Yellowstone Park Reference Library.

Lott, Melvina J. "Diary of a Yellowstone Park Trip, 1880." Typed MS. Montana State University Library, Bozeman.

Ludlow, William. *Report of a Reconnaissance from Carroll, Montana Territory, on the Upper Missouri, to the Yellowstone National Park, and Return, Made in the Summer of 1875*. Washington, 1876.

Maguire, H. N. "The New Eldorado." *Avant Courier*. Bozeman, MT, January 7, 1876.

Mallon, Henry, et al. Interviewed by A. L. Haines at Mammoth Hot Springs, July 5, 1961. Audio Tape 61–2. Yellowstone Park Reference Library.

Malone, Michael P., and Richard B. Roeder. *Montana: A History of Two Centuries*. Seattle, 1976.

Manton, Margaret. "In the Yellowstone." *The Chicago Tribune*, December 6, 1891.

Marler, George D. [Hot Springs of the Norris Geyser Basin]. Mimeographed, [1962]. Yellowstone Park Reference Library.

———. *Studies of Geysers and Hot Springs Along the Firehole River, Yellowstone National Park*. Yellowstone Library and Museum Association, 1964.

Marshall, George W. Biographical Statement from the Bancroft Library [1885]. Photostat. Yellowstone National Park Library.

Mattes, Merrill J. *Colter's Hell and Jackson's Hole*. Yellowstone National Park, 1962.

Matthews, Albert. "The Word Park in the United States." *Publications of the Colonial Society of Massachusets*. April 1904.

Matthews, L. F. [Letter to manager J. H. Dean, October 1, 1898]. Northern Pacific Railroad Files, President, 210-A.

McCartney, James C. Statement of James C. McCartney, May 15, 1891. Yellowstone Archives, Doc. 1137.

McGuirk, Matthew. Statement of Matthew McGuirk, May 5, 1891. Yellowstone Archives., Doc. 1149.

McIntyre, Robert N. "Memorial Plaque for the new Chittenden Bridge." Memorandum to the Programs Officer, Yellowstone National Park. June 6, 1961.

Mattoon, A. M. "The Yellowstone National Park, Summer of 1889." Handwritten MS, 1907. Yellowstone Park Reference Library.

Miller, Martin F. *West Thumb Geyser Basin*. Map drawn July 1960.

Mitchell, S. Weir. "Through the Yellowstone Park to Fort Custer." *Lippincott's Magazine*. June-July 1880.

Montana Assembly. *Acts, Resolutions and Memorials of the Territory of Montana Passed by the First Legislative Assembly, Convened at Bannack, December 12, 1864*. Virginia City, MT, 1866.

Moyle, Don. "Crow Agency Man Relates Custer Scout's Story of Battle of the Little Big Horn." *Billings Gazette*, MT. June 21, 1953.

Murri, Ernest. Interview with John Douglas and Garth Nelson, at Ashton, Idaho. May 5, 1968.

National Park Service. *Yellowstone National Park*. Washington, 1923.

Nettleton, A. B. Letter to F. V. Hayden, October 27, 1871. National Archives, Records of the Department of Interior, Geological Survey, Letters received by F. V. Hayden, 1871.

Norris, Philetus W. *Report Upon the Yellowstone National Park, to the Secretary of the Interior, for the Year 1877* (dated October 20, 1877). Washington, 1877.

———. Letter to the Secretary of the Interior, February 11, 1878. United States, 45th Cong., 2nd Sess. House. Ex. Doc. No. 75, April 11, 1878.

———. *Report Upon the Yellowstone National Park, to the Secretary of the Interior, for the Year 1878.* Washington, 1879.

———. *Report Upon the Yellowstone National Park to the Secretary of the Interior, for the Year 1879.* Washington, 1880.

———. *Annual Report of the Superintendent of the Yellowstone National Park to the Secretary of the Interior for the Year 1880.* Washington, 1881.

———. *Fifth Annual Report of the Superintendent of the Yellowstone National Park,* (dated December 1, 1881). Washington, 1881.

———. *The Calumet of the Coteau Together with a Guide-book of the Yellowstone Park.* Philadelphia, 1884.

Norton, Harry J. *Wonderland Illustrated; or, Horseback Rides Through the Yellowstone National Park.* Virginia City, MT, 1873.

Orth, Donald J. Memorandum, "Proposing Names for Unnamed Domestic Features," March 8, 1877. Copy forwarded to Yellowstone National Park, November 27, 1979.

Osmond, Mabel C. "Memories of a Trip Through Yellowstone Park in 1874," Typescript [1927]. Yellowstone Park Reference Library.

Panton, S. P. "Early Days in Yellowstone Recalled by N. P. Surveyor." *Wolf Point News,* MT, November 23, 1939.

Parkinson, Edward S. *Wonderland; or, Twelve Weeks in and out of the United States.* Trenton, NJ, 1894.

Paul, Caroline L. "Notes on Yellowstone National Park," July 24–30, 1897. Excerpted from the account of a Christian Endeavor excursion from Massachusetts to Yosemite National Park, June 28–August 4, 1897. [Copy made at Yosemite, November 30, 1948].

Payette, B. C. *The Oregon Country Under the Union Jack.* Montreal, Canada, 1962.

Peale, A. C. "A. C. Peale U.S. Geological Survey, 1872." MS diary. Yellowstone Park Reference Library.

Peek, Ernest D. "Report upon the Construction, Repair, and Maintenance of Roads and Bridges in the Yellowstone National Park," *Annual Report of the Chief of Engineers, 1908.* Washington, 1908.

Pitcher, John. *Report of the Acting Superintendent of the Yellowstone National Park to the Secretary of the Interior, 1901.* Washington, 1901.

———. *Report of the Acting Superintendent of the Yellowstone National Park to the Secretary of the Interior, 1902.* Washington, 1902.

———. Letter to John Yancey, May 31, 1902. LS 11. Yellowstone National Park Archives.

Pitcher, John. *Report of the Acting Superintendent of the Yellowstone. National Park to the Secretary of the Interior, 1903.* Washington, 1903.

———. *Report of the Acting Superintendent of the Yellowstone National Park to the Secretary of the Interior, 1904.* Washington, 1904.

———. *Report of the Acting Superintendent of the Yellowstone National Park to the Secretary of the Interior, 1905. Washington, 1905.*

Potter, Lena. Interviewed by A. L. Haines at Gardiner, Montana, April 20, 1962. Audio tape 61–5. Yellowstone Park Reference Library.

Potts, Daniel T. Letter to Robert T. Potts, July 8, 1827. MS in Yellowstone Park Reference Library.

Powell, J. W. *Fifth Annual Report* [of the United States Geological Survey] *1883–84.* Washington, 1885.
———. *Sixth Annual Report* [of the United States Geological Suvey] *1884–85.* Washington, 1885.
———. *Seventh Annual Report* [of the United States Geological Survey] *1885–86,* 4 Vols. Washington, 1888.
———. *Eighth Annual Report* [of the United States Geological Survey] *1886–87,* 2 Parts. Washington, 1888.
Quaife, M. M. *Yellowstone Kelly.* Lincoln, NE, 1973.
Raftery, John H. *The Story of Yellowstone Park.* Butte, MT, 1912.
———. *A Miracle in Hotel Building.* Yellowstone Park Company, [1912].
Rand McNally. *Map of Montana,* c. 1885.
Raymond, Rossiter W. "Sketches of Western Travel." *New North-West.* Deer Lodge, MT, May 25, 1872.
Raynolds, William F. *U.S. War Department Map of the Yellowstone and Missouri Rivers and their Tributaries explored by Capt. W. F. Raynolds Topl. Engrs., and 1st Lt. H. E. Maynardier 10th Infy, Assistant, 1859–60.* National Archives, RG–77, Records of the Office of the Chief of Engineers, Headquarters map file, Q 106–1.
———. *The Report of Brevet Brigadier General W. F. Raynolds on the Exploration of the Yellowstone and the Country Drained by That River.* United States. 40th Cong., 1st Sess. Senate. Ex. Doc. No. 77 (1868).
Reeves, Mary. [Three letters concerning a trip through Yellowstone National Park in 1889]. Originals in Yellowstone Park Reference Library.
Replogle, Wayne F. *Yellowstone's Bannock Indian Trails.* Yellowstone Park, WY, 1956.
———. "Rip' Rap." *Yellowstone News.* August 7, 1975.
Richardson, James. *Wonders of the Yellowstone.* New York, 1873.
Riley, W. C. *Grand Tour Guide to the Yellowstone National Park.* St. Paul, MN, 1889.
Ritchie, Edith. Interviewed by A. L. Haines at Gardiner, Montana, November 7, 1961. Audio tape 61–3. Yellowstone Park Reference Library.
Roberts, Edward. *Shoshone and Other Western Wonders.* New York 1888.
Rogers, Edmund B. Letter To Horace M. Albright, May 27, 1938. Copy in author's file.
———. Memorandum for the Director, National Park Service, October 16, 1944. Yellowstone National Park.
———. *History of Legislation Relating to the National Park System Through the 82d Congress.* Denver 1958.
Rogers, H. D., and A. Keith Johnnson. *Territories of Washington and Oregon* (Map). London, U.K., 1857.
Roosevelt, Theodore. Presidential Executive Order, May 22, 1902.
Ross, Alexander. *The Fur Hunters of the Far West; A Narrative of Adventures in the Oregon and Rocky Mountains,* 2 Vols. London, U. K., 1855.
Russell, Osborne. *Journal of a Trapper.* Ed. by Aubrey L. Haines. Portland, OR, 1955.
Samuell, Yellowstone C. "Heavers, Hashers, Skinners & Dudes." *Montana Magazine of History,* Jan. 1980.
Sanford, James C. "Construction of Military Road from Fort Washakie to Mouth of Buffalo Fork of Snake River, Wyoming." *Annual Report of the Chief of Engineers for 1899.* Washington, 1899.

Savage, Joseph. "The Wonders of the Yellowstone." *Western Home Journal.* Lawrence, KS, September 19, 1872.
Scales, Bill. Letter to the Chief Ranger, Yellowstone National Park, May 20, 1963.
Schmide, Carl E. *A Western Trip.* Detroit, MI, 1910.
Scrapbook No. 1. [Clippings from early newspapers — mostly unidentified or undated.] Yellowstone Park Reference Library.
Sessions, Francis C. "Yellowstone National Park." *Magazine of Western History.* September 1887.
Shankland, Robert. *Steve Mather of the National Parks.* New York, 1951.
Sheridan, Philip H. *Report of Lieut. General P. H. Sheridan, dated September 20, 1881, of His Expedition Through the Big Horn Mountains, Yellowstone National Park, etc.* Washington, 1882.
Sheridan, Philip H., et al. "Report of Col. O. M. Poe, U.S. Engineers." *Reports of Inspections Made in the Summer of 1877.* Washington, 1878.
Sheridan, Philip H., et al. *Report of an Exploration of Parts of Wyoming, Idaho and Montana, in August and September, 1882.* Washington, 1882.
Sherman, Thomas E. "Across the Continent. II — The National Park." *Woodstock Letters,* Vol. XI, 1882.
[Shively, John]. "Life With the Nez Perce." *New North-West.* Deer Lodge, MT, September 14, 1877.
Silliman, Lee. "As Kind and Generous a Host as Ever Lived." *American West.* July-August, 1979.
Slough, John P. Letters to Superintendent Horace M. Albright, August 6, 1922 and January 13, 1924. Yellowstone Park Reference Library.
Smalley, E. V. "The New Northwest." *Century.* September 1882.
———, compiler. *Northern Pacific Railroad; Book of Reference, for Use of the Directors and Officers of the Company.* New York, 1883.
Stanley, Edwin J. *Rambles in Wonderland, or, Up the Yellowstone and Among the Geysers and Curiosities of the National Park.* New York, 1878.
Steele, David M. *Going Abroad Overland.* New York, 1917.
Stilson, Gardner S. *A Tale of the Yellowstone; or, In a Wagon Through Western Wyoming and Wonderland.* Jefferson, IA, 1901.
Stout, Tom. *Montana, Its Story and Biography,* 3 Vols. Chicago, 1921.
Strahorn, Carrie A. *Fifteen Thousand Miles by Stage,* 2nd Ed. New York, 1915.
Strahorn, Robert E. *To the Rockies and Beyond.* Chicago, 1881.
Strong, William E. *A Trip to the Yellowstone National Park in July, August and September 1875.* Washington, 1876.
Stuart, Granville. *Montana As It Is.* New York, 1865.
Superintendent's Monthly Reports. Yellowstone Park Reference Library.
Surveyor General of Montana. *Map of the Territory of Montana to Accompany the Report of the Surveyor General [Meredith], 1869.* National Archives. RG–49, Records of the Bureau of Land Management, Old map file, Mont. 2.
———. *Map of the Territory of Montana to Accompany the Report of the Surveyor General [Washburn], 1869.* National Archives. RG–49, Records of the Bureau of Land Management, Old map file, Mont.3.
Switzer, Malthus A. "Diary of a Trip to the Geysers in 1876." MS. Montana State University Library, Bozeman.

Synge, Georgina M. *A Ride Through Wonderland.* London, U.K., 1892.
Thomas, Thomas H. "Yellowstone Park Illustrated," Part I. *The Graphic.* August 11, 1888.
Thompson, J. H. "Memorandun of a Trip from Fort Ellis, MT, to the Yellowstone Park, Sept. 1882." MS diary. Yellowstone Park Reference Library.
Thwaites, Reuben G. *Original Journals of Lewis & Clark Expedition, 1804–1806,* Vol. II. New York, 1904.
Toll, Roger W. [Annual Report for Yellowstone National Park, 1929]. Typed.
———. "Annual Report For Yellowstone National Park, 1931." Mimeographed.
———. "Annual Report for Yellowstone National Park, 1932." Mimeographed.
———. "Annual Report for Yellowstone National Park, 1933." Mimeographed.
Topping, Eugene S. *The Chronicles of the Yellowstone.* St. Paul, MN, 1883.
Trumbull, Walter. "Yellowstone Papers." *Weekly Rocky Mountain Gazette.* Helena, MT, October 3, 1870.
———. "Yellowstone Papers." *Weekly Rocky Mountain Gazette.* Helena, MT, October 24, 1870.
———. "Yellowstone Papers." *Weekly Rocky Mountain Gazette.* Helena, MT, October 31, 1870.
———. "Yellowstone Papers — No. 1." *Daily Rocky Mountain Gazette.* Helena, MT, October 18, 1870.
———. "Yellowstone Papers — No. 2." *Daily Rocky Mountain Gazette.* Helena, MT, October 19, 1870.
———. "The Washburn Yellowstone Expedition, No. 1." *Overland Monthly.* May, 1871.
———."The Washburn Yellowstone Expedition, No. 2." *Overland Monthly.* June, 1871.
Turpin, Francis L. "A Trip Through Yellowstone Park, 1895." Typescript prepared by Florence Ballinger. Montana State University Library, Bozeman.
Turrill, Gardner S. *A Tale of the Yellowstone; or, In a Wagon Through Wyoming and Wonderland.* Jefferson, IA, 1901.
United States Army. "Record Book of Interments, Fort Yellowstone, Wyo." Yellowstone Archives, Doc. 155. Yellowstone Park Reference Library.
United States Board on Geographic Names. *Decisions of the United States Board on Geographic Names, July 1, 1936 to June 2, 1937.* Washington, 1937.
———. *Decisions of the United States Board on Geographic Names.* List 6202. Washington, 1962.
———. *Decisions of the United States Board on Geographic Names.* List 6602. Washington, 1966.
United States Board on Geographic Names. [Decisions]. List 8102. Typescript, 1981. Yellowstone Park Reference Library.
United States Engineer Department. [Route maps of Yellowstone Park roads, 1891.]. 26 blueprints. Yellowstone National Park.
United States Forest Service. *Absaroka National Forest, Montana* (base map). Washington, 1937.
———. *Custer National Forest (Beartooth Division), Montana* (base map). Washington, 1938.
———. *Gallatin National Forest, Montana* (base map). Missoula, MT, 1947.

———. *Gallatin National Forest, Hebgen Lake Ranger District, 1971* (base map). Forest Series.
———. *Shoshone National Forest, Wyoming* (base map). Washington, 1927.
———. *Shoshone National Forest, Wyoming* (base map). Washington, 1940.
———. *Targhee National Forest, Idaho and Wyoming,* (base map). Washington, 1925.
———. *Targhee National Forest, Idaho and Wyoming,* (base map). Ogden, UT, 1944.
———. *Targhee and Teton National Forests, Idaho and Wyoming, 1936.* Recreation map.
———. *Teton National Forest, Wyoming,* (base map). Washington, 1930.
———. *Teton National Forest, Wyoming,* (base map). Washington, 1938.
———. *Teton National Forest, Wyoming,* (base map). Washington, 1944.
United States Geographic Board. *Decisions of the United States Geographic Board, Yellowstone National Park, Wyoming.* May 7, 1930. Washington, 1930.
———. *Sixth Report of the United States Geographic Board, 1890–1932.* Washington, 1933.
United States Geological Survey. *Bulletin 47.* Washington, 1884.
———. *Sixth Annual Report, for the Fiscal Year Ending June 30, 1885.* Washington, 1886.
———. *Yellowstone National Park, Canyon Sheet,* N4430–4500/W11000–11030. Edition of 1886.
———. *Yellowstone National Park, Gallatin Sheet,* N4430–4500/W11030–11100. Edition of 1886.
———. *Yellowstone National Park, Lake Sheet,* N4400–4430/W11000–11030. Edition of 1886.
———. *Yellowstone National Park, Shoshone Sheet,* N4400–4430/W11030–11100. Edition of 1886.
———. *Yellowstone National Park and North Western Wyoming, 1886,* N4400–4500/W11000–11100. New York, 1886.
———. *Eighth Annual Report, 1886–1887.* Washington, 1888.
———. *Ninth Annual Report, for the Fiscal Year Ending June 30, 1888, 4 Vols.* Washington, 1889.
———. *Montana, Livingston Sheet,* N4500–4600/W11000–11100. Edition of 1891.
———. *Geologic Atlas of the United States, Livingston Folio, Mont..* Washington, 1894.
———. *Geologic Atlas of the United States, Three Forks Folio, Mont.* Washington, 1895.
———. *Geologic Atlas of the United States, Yellowstone National Park Folio, Wyoming,* 4 Sheets (Canyon, Gallatin, Lake, Shoshone). Washington, 1896.
———. *Geologic Atlas of the United States, Absaroka Folio, Wyo.* Washington, 1899.
———. *Grand Teton National Park, Wyoming,* 1:125,000. Washington, 1901, reprinted 1923 & revised 1929.
———. "Detail Map of Mammoth Hot Springs Travertine Terraces," 1:2400. Sheet XX, *Yellowstone Monograph XXXII.* New York, 1904.
———. "Mammoth Hot Springs and Vicinity," 1:14400. Sheet XVIII, *Yellowstone Monograph XXXII.* New York, 1904.
———. "Norris Geyser Basin," 1 in. = 300 ft. Sheet XXI, *Yellowstone Monograph XXXII.* New York, 1904.
———. "Lower Geyser Basin," 1 in. = 800 ft. Sheet XXII, *Yellowstone Monograph XXXII.* New York, 1904.

———. "Egeria Springs," (Midway Geyser Basin), 1 in. = 200 ft. Sheet XXIII, *Yellowstone Monograph XXXII.* New York, 1904.
———. "Upper Geyser Basin," 1 in. = 500 ft. Sheet XXIV, *Yellowstone Monograph XXXII.* New York, 1904.
———. "Shoshone Geyser Basin," 1 in. = 225 ft. Sheet XXVI, *Yellowstone Monograph XXXII.* New York, 1904.
———. "Yellowstone National Park and Portions of Yellowstone Forest Reserve," 1:250,000. Sheet XVI, *Yellowstone Monograph XXXII.* New York, 1904.
———. *Shoshone, Wyo.*. N4400-W11030/30. 1911.
———. *Administrative Map of Yellowstone National Park, Wyo.-Mont.-Ida.*, 1:125,000. 1912.
———. *Lake, Wyo.*. N4400-W11000/30. 1911, re 1922.
———. *Shoshone, Wyo.*. N4400-W11030/30. 1911, re 1927.
———. *Gallatin, Wyo.*. N4430-W11030/30. 1911, re 1928.
———. *Yellowstone National Park, Wyo.-Mont.-Ida.*, 1:125,000. 1912, re 1928.
———. *Yellowstone National Park, Wyo.-Mont.-Ida.*, 1:125,000. 1930.
———. *Mt. Leidy, Wyo.*. N4330-W11000/30. 1902, re 1932.
———. *Cutoff Mtn., Mont.-Wyo.*. N4500-W11000/15. 1942.
———. *Mt. Wallace, Mont.*. N4500-W11015/15. 1943.
———. *Deep Lake, Wyo.*. N4445-W10915/15. 1950.
———. *Bozeman, Mont.*. N4530-W11100/15. 1953.
———. *Yellowstone National Park, Wyo.-Mont.-Ida.*. 1:125,000. 1930, re 1954.
———. *Crown Butte, Mont.* N4500-W11100/15. 1955.
———. *Gardiner, Mont.* N4500-W11030/15. 1955.
———. *Miner, Mont.-Wyo.* N4500-W11100/15. 1955.
———. *Beartooth Butte, Wyo.* N4445-W10930/15. 1956.
———. *Grassy Lake Reservoir, Wyo.* N4400-W11045/15. 1956.
———. *Huckleberry Mountain, Wyo.* N4400-W11030/15. 1956.
———. *Old Faithful, Wyo.* N4415-W11045/15. 1956.
———. *Pilot Peak, Wyo.* N4445-W10945/15. 1956.
———. *Sunlight Peak, Wyo.* N4430-W10945/15. 1956.
———. *West Thumb, Wyo.* N4415-W11030/15. 1956.
———. *Buffalo Lake, Idaho-Wyo.-Mont.* N4415-W11100/15. 1957.
———. *Warm River Butte, Idaho-Wyo.* N4400-W11100/15. 1957.
———. *Madison Junction, Wyo.* N4430-W11045/15. 1958.
———. *Mammoth, Wyo.-Mont.* N4445-W11030/15. 1958.
———. *Mount Holmes, Wyo.-Mont.*. N4445-W11045/15. 1958.
———. *Norris Junction, Wyo.* N4430-W11030/15. 1958.
———. *Tepee Creek, Mont.-Wyo.* N4445-W11100/15. 1958.
———. *West Yellowstone, Mont.-Wyo.-Idaho.* N4430-W11100/15. 1958.
———. *Abiathar Peak, Wyo.-Mont.* N4445-W11000/15. 1959.
———. *Canyon Village, Wyo.* N4430-W11015/15 1959.
———. *Eagle Peak, Wyo.* N4415-W11000/15. 1959.
———. *Frank Island, Wyo.* N4415-W11015/15. 1959.
———. *Mount Hancock, Wyo.* N4400-W11015/15. 1959.
———. *Pelican Cone, Wyo.* N4430-W11000/15. 1959.
———. *Tower Junction, Wyo.-Mont.* N4445-W11015/15. 1959.

———. *Two Ocean Pass, Wyo.* N4400-W11000/15. 1959.
———. *Yellowstone National Park, Wyo.-Mont.-Idaho.* N4408-W10948/59x82. 1961.
United States Geological Survey of the Territories. *Gibbon Geyser Basin* (Norris). Gannett/Mushback, 1878. Yellowstone Park Reference Library.
———. *Lower Geyser Basin.* Gannett/Mushback, 1878. Yellowstone Park Reference Library.
———. *Egeria Springs* (Midway), Firehole River. Gannett/Mushback, 1878. Yellowstone Park Reference Library.
———. *Upper Geyser Basin.* Gannett/Mushback, 1878. Yellowstone Park Reference Library.
———. *Shoshone Geyser Basin,* west end of Shoshone Lake. Gannett/Mushback, 1878. Yellowstone Park Reference Library.
———. *Hot Springs on Witch Creek Near Heart Lake.* Gannett/Mushback, 1878. Yellowstone Park Reference Library.
———. *Yellowstone National Park,* an incomplete topographic map at the scale of 1:126,720. Surveyed in 1878, published in 1881. Yellowstone Park Reference Library.
United States Public Roads Administration. "West Thumb Hot Spring Area," Scale, 1 in. = 100 ft.. F.A.L., April 1941.
*United States Statutes at Large.* Vol. 17, Chapt. 24.
*United States Statutes at Large.* Vol. 28, Chapt. 72.
Vinton, Stallo. *John Colter, Discoverer of Yellowstone Park.* New York, 1926.
Washburn, Henry D. *Map of the Public Land Surveys in Montana Territory to Accompany the Annual Report of the Commissioners of the General Land Office, 1869.* National Archives, RG–49, GLO(OMF) — Montana 3.
———. "The Yellowstone Expedition." *Daily Herald.* Helena, MT, September 28, 1870.
———. *Route of Washburn's party 1870, being a tracing of map made to accompany Washburn's report to Dept. of Interior.* National Archives, RG–49 GLO(OMF) — Montana 5.
Watson, W. Verde. MS map of Norris Geyser Basin, 1954. Yellowstone Park Reference Library.
[Weaver, David B., from "D. W."]. "The Upper Yellowstone" *Montana Post.* Virginia City, MT, August 31, 1867.
Weaver, David B. "Early Days in Emigrant Gulch." *Contributions to the Historical Society of Montana,* VII. Helena, MT, 1910.
Weber, Nan. Culver Papers made available to the author, January 11, 1990. Salt Lake City, UT.
Weimer, Josiah W. Letter to his family, Butler County, KS, October 16, 1883. Yellowstone Park Reference Library.
Wheat, Carl I. *Mapping the Transmississippi West, 1540–1861,* I. San Francisco, 1957.
Wheeler, Olin D. "The Late James Gemmell." *Contributions to the Historical Society of Montana,* II. Helena, MT, 1896.
———. "Yellowstone Park." *6,000 Miles Through Wonderland.* St. Paul, MN, 1893.
———. "Yellowstone National Park." *Wonderland '96.* St. Paul, MN, 1896.
———. "The Yellowstone Wonderland." *Wonderland 1905.* St. Paul, MN, 1905.
White, Alma. *With God in Yellowstone.* Zarepeth, NJ, 1920.

Whithorn, Doris. "A Grizzly Got Frenchy." Typescript, 1963.
Whitmell, Charles T. "The American Wonderland: The Yellowstone National Park." *Report and Transactions of the Cardiff Naturalist's Society* (1885). Cardiff, Wales, 1886.
Whitney, Josiah D. *The Yosemite Book.* 1868.
Whittlesey, Lee H. *Yellowstone Place Names.* Helena, MT, 1988.
Wilkes, Charles. *Map of the Oregon Territory.* U.S. Exploring Expedition. 1841.
Wingate, Geo. W. *Through the Yellowstone Park on Horseback.* New York, 1886.
Wister, Owen. "Old Yellowstone Days." *Harpers Monthly.* March 1936.
Wunderlich, Rudolf. "Recent Acquisitions in Important Western and Sporting Paintings and Sculptures." *The Kennedy Quarterly.* October 1962.
Wylie, William W. *Yellowstone National Park; Or the Great American Wonderland.* Kansas City, MO, 1882.
Wyoming Laws. *Laws Appertaining to the Yellowstone National Park, Including Statutes of the Territory of Wyoming Approved Mar. 6, 1884.* Cheyenne, W.T., 1884.
Yeager, Dorr G. "A Reminder of Early Days." *Yellowstone Nature Notes.* April 1929.
Yellowstone Library and Museum Association. "Mud Volcano Trail" (pamphlet). 1966.
Yellowstone National Park. "Ranger Naturalist Manual." Mimeographed, 1927.
Yellowstone National Park. "Ranger Naturalist Manual." Mimeographed, 1928.
Yellowstone National Park. Press release from the office of the Park Superintendent following the death of Lord Dunraven. January, 1927.
*Yellowstone National Park, Wyoming, 1923* (circular). Yellowstone Park Reference Library.
Yellowstone Park Company. Structural plans deposited with the Montana Historical Society, Helena.

# INDEX

Yellowstone Park place names are listed in **bold** type if approved and in *italic* type if obsolete or otherwise unapproved. Parenthetical tags refer to the correct usage or the general location where no approved form is known.